STOCHASTIC SCHEDULING

Stochastic scheduling occurs in the area of production scheduling. There is a dearth of work that analyzes the variability of schedules. In a stochastic environment, in which the processing time of a job is not known with certainty, a schedule typically is analyzed based on the expected value of a performance measure. This book addresses this problem and presents algorithms to determine the variability of a schedule under various machine configurations and objective functions. It is intended for graduate and advanced undergraduate students in manufacturing, operations management, applied mathematics, and computer science, and it is also a good reference book for practitioners. Computer software containing the algorithms is also provided on an accompanying Web site (www.cambridge.org/sarin) for ease of student and user implementation.

Dr. Subhash C. Sarin earned a Ph.D. at North Carolina State University in 1978. He has made research contributions in production scheduling, sequencing, applied mathematical programming, and analyzing and designing algorithms for the operational control of manufacturing systems. Dr. Sarin is the Paul T. Norton Endowed Professor at Virginia Polytechnic Institute and State University's Grado Department of Industrial and Systems Engineering.

Balaji Nagarajan is a graduate of the Grado Department of Industrial and Systems Engineering at Virginia Tech. He currently works in the aviation industry developing optimization models and solutions for various airline operations. His research interests include mathematical modeling, optimization, and their applications in practice.

Lingrui Liao is a doctoral student in the Grado Department of Industrial and Systems Engineering. His research interests lie in the areas of production planning and scheduling, and applied mathematical programming.

Stochastic Scheduling

Subhash C. Sarin
Balaji Nagarajan
Lingrui Liao
Virginia Polytechnic Institute and State University

CAMBRIDGE UNIVERSITY PRESS
Cambridge, New York, Melbourne, Madrid, Cape Town, Singapore,
São Paulo, Delhi, Dubai, Tokyo

Cambridge University Press
32 Avenue of the Americas, New York, NY 10013-2473, USA

www.cambridge.org
Information on this title: www.cambridge.org/9780521518512

First published 2010

Printed in the United States of America

A catalog record for this publication is available from the British Library.

Library of Congress Cataloging in Publication data
Sarin, Subhash Chander.
  Stochastic scheduling / by Subhash C. Sarin, Balaji Nagarajan, Lingrui Liao.
    p. cm.
  Includes bibliographical references and index.
  ISBN 978-0-521-51851-2
  1. Production scheduling. I. Nagarajan, Balaji. II. Liao, Lingrui. III. Title.
  TS157.5.S27 2010
  658.5′3–dc22   2010000032

ISBN 978-0-521-51851-2 Hardback

Additional resources for this publication at www.cambridge.org/sarin.

Contents

Foreword

Over the years, I have taught introductory as well as advanced courses in job scheduling. The problem is fascinating in its variety, which is truly astounding, and in the methodologies that have been offered by many researchers for its resolution. But I have always felt let down when it came to job scheduling under uncertainty: The absence of a comprehensive treatment of these real-life scenarios was a handicap that I felt acutely. After reading this book, I do not feel that way any more.

The fixation on "averages" is often meaningless, if not downright misleading. True, while the statement "on average, such and such phenomenon behaves in a certain way" or "has a particular value" is often a useful bit of information (one is usually interested in the average water temperature at the seaside before taking a jump into the ocean), it often carries insufficient information for intelligent decision making (knowing that one's heartbeat is good "on average" does not help to diagnose the ailment if sometimes it stops beating altogether!). Here is where the "range of variation," as measured by the variance (or any other measure of dispersion), becomes invaluable.

Scheduling under stochastic conditions has received scant attention from authors in the field. Usually it appears, if at all, in the form of one or two (thin) chapters that introduce the problem and give a treatment based mostly on the assumption that the processing times of the jobs are exponentially distributed. The feeling always has persisted: Whoever heard of processing times that are exponentially distributed? And how to behave if they are not?

Heaven knows (also researchers in the field know) that scheduling under conditions of uncertainty with the objective of optimizing the expected value of the adopted criterion, be it job-focused or time-focused, is difficult enough. But adding another dimension on top of it in the form of the variance of the declared criterion seems like adding insult to injury and stretching credibility in the analyst's ability to come up with an "answer" – any answer! This is what this book is all about. Sarin makes no attempt to hide such difficulty – on the contrary, for each scenario he presents the mathematical model (if one exists) in all its glorious complexity and then proceeds to discuss ways and means of approaching a solution through branch and bound or heuristics.

To my utter delight, throughout this monograph I discovered a few concepts that have general applicability well beyond the subject matter of the book. For instance, in Chapter 8, the reader is introduced to the concept of finite mixture of distributions, in which a random variable is represented as a convex combination of a finite set of random variables. Usually, the component random variables are assumed to be normally distributed, but they can be any other distribution. Sarin is meticulous in explaining the approach and citing the appropriate references. Anyone interested in "fitting a continuous distribution to a given set of data" or in "approximating a distribution by a set of normal distributions" would welcome the discussion. I particularly enjoyed being treated to the rare view of approximating the *uniform* distribution with a convex mixture of *normal* distributions as the number of iterations of the approximation increases from 1 to 113,179. The further application of these concepts to the approximation of the distribution of the completion time of a project (in which the "jobs" are further constrained by precedence relations) provides an excellent introduction to the myriad problems faced by analysts in treating project-related problems.

A pleasant surprise to the owner of this book is the software XVA-Sched (for the *e*xpectation-*v*ariance *a*nalysis of a *sched*ule) that accompanies it, as well as the instructions on how to use it. This software helps the user to implement the methodologies developed in previous chapters to determine the expectation-variance-efficient schedules.

Subhash C. Sarin has written a gem of a research monograph that shall find its way to the shelf of each researcher and worker in the field of scheduling. It is long overdue. And I feel privileged in having been given a peek at its contents before its publication.

Salah E. Elmaghraby
University Professor
North Carolina State University
Raleigh, North Carolina, USA
July 13, 2009

Preface

Scheduling is a decision-making process that is commonly encountered in practice – in both production and nonproduction-related environments. The effectiveness of a schedule depends on how well it performs in the environment for which it is designed. For an environment that involves a stochastic element, which is more often the case in practice, a typical approach presented in the literature has been to determine a schedule that optimizes the expected value of a performance measure. However, optimizing merely the expectation of a performance measure for a schedule is not enough for use in such an environment because variability of the measure plays an important role in influencing the performance of the schedule. It is, therefore, essential that a schedule be selected for implementation that has suitable values for both the expectation and variance of the performance measure in consideration. In this book, we present methodologies for determining such a schedule, and in this respect, this book is different from the other books on scheduling.

The book is organized as follows. In Chapter 1, we focus on the impact of uncertainty (variability) in scheduling, and the need for efficient modeling of stochastic scheduling problems and for devising effective scheduling strategies to counter the impact of variability. We specifically highlight the prevalence of variability in job processing times and elicit the issue of neglecting variance in schedule optimization and the significance of considering variance. Furthermore, we enunciate the need for a comprehensive analytical evaluation of the expectation and variance of different performance measures. Chapters 2 and 3 review methods presented in the literature to address the issue of variability of a performance measure for a schedule. In particular, Chapter 2 deals with robust scheduling approaches to hedge against processing time variability. Various model formulations and solution methodologies are presented in detail. Chapter 3 deals with another approach that determines a set of "nondominating schedules" or "expectation-variance efficient schedules" and selects a preferred schedule from such a set. We present our work in Chapters 4, 5, 6, 7, and 8 wherein a comprehensive analysis of schedules for various scheduling environments and different performance measures is presented. Chapters 4, 5, 6, and 7, respectively, deal with scheduling in a single-machine, flow-shops, job-shops, and parallel-machines environments. We develop closed-form expressions (wherever possible, and devise methodologies otherwise) to evaluate the expectation and variance of various performance measures for a schedule. The methodologies are illustrated using example problems. The closed-form expressions for many performance measures rely on the assumption that job processing times follow normal distributions. This assumption is relaxed in Chapter 8, and we consider the case of general processing time distributions for jobs. We use a finite mixture model to represent a given processing time distribution of a job by a convex combination of normal distributions. Subsequently, we use the methodologies presented in previous chapters to perform the expectation-variance analysis of a schedule for each of the environments presented in those chapters whose analysis relies on the assumption of normal distribution. Also, we demonstrate, in this chapter, the use of a finite mixture model to perform the expectation-variance analysis for the completion time of an activity network.

An accompanying Web site (www.cambridge.org/sarin) contains a software package to help implement the methodologies developed in Chapters 4, 5, 6, 7, and 8. This is user-friendly software termed XVA-Sched (for the *ex*pectation-*va*riance *a*nalysis of a *sched*ule). The instructions to use the software are included in an appendix.

The material presented in the book can be used as a supplement to a course in sequencing and scheduling, and to courses in related areas at both graduate and advanced undergraduate levels. As background, it requires mathematical maturity and introductory knowledge of probability theory and optimization concepts and methodologies. The book provides useful ideas and algorithms for practitioners, and it can serve as a useful research reference.

My first and foremost thanks go to my graduate students Balaji Nagarajan and Lingrui Liao. For their many direct contributions, I consider them co-authors of this book. I would also like to extend my sincere thanks to the anonymous reviewers for their careful reading of the manuscript and insightful comments.

A project of this magnitude cannot be accomplished without the unconditional support, encouragement, and love of the family. For this, I would like to thank my wife, Veena, and our sons, Sumeet and Shivan.

Subhash C. Sarin
Blacksburg, Virginia
August 14, 2009

STOCHASTIC SCHEDULING

1 Introduction

1.1 Uncertainty

The realm of manufacturing is replete with instances of uncertainties in job processing times, machine statuses (up or down), demand fluctuations, due dates of jobs, and job priorities. These uncertainties stem from the inability to predict with sufficient accuracy information about product demand, job processing times, and occurrences such as unexpected machine breakdowns and arrivals/cancellations of orders. Although highly efficient forecasting methods are currently available, product demand errors invariably occur in the production system. Process variability is another significant factor that introduces variations and uncertainty into the manufacturing process.

Uncertainty is inarguably an undesirable factor in the manufacturing process because it does not give production managers complete control over the manufacturing process. Some of the ill effects of these uncertainties include system instability, excess inventory, customer dissatisfaction because due dates are not met, and, more important, loss of revenue. Recent advances in manufacturing management techniques, such as agile manufacturing, have made variability an important design criterion in order to ensure predictability and dependability of production systems. *Agility* is the ability of a system to thrive and prosper in an environment of constant and unpredictable change.

1.2 Uncertainty in Scheduling

As true as it would be with any other field within manufacturing, the uncertainty factor is of considerable importance in production scheduling. Scheduling is a decision-making process that plays an important role in most manufacturing as well as in most information-processing environments. From a manufacturing perspective, a *scheduling problem* is primarily the determination of the starting times of the jobs waiting to be processed, on a single machine or multiple machines (resources) for the objective of optimizing an appropriate performance measure of interest. The randomness in the scheduling system could be due to varying processing times, machine breakdowns, and incomplete information about customer due dates, among other things. *Deterministic scheduling* involves solving a scheduling problem with the objective of optimizing a performance measure of interest when the various parameters, *viz.,* job processing times, due dates, release dates, and so on, are known with certainty. On the other hand, *stochastic scheduling* deals with problems when at least one of these parameters is not known with certainty. Scheduling under stochasticity is relatively more complex and difficult than its deterministic counterpart.

1.3 Modeling Uncertainty in Scheduling

As stated earlier, uncertainty has a major impact on scheduling decisions. Conventionally, in stochastic scheduling, the uncertain or variable scheduling parameters are modeled as random variables, and researchers endeavor to optimize a performance measure of interest that is suitable to the problem at hand. In a majority of the work, the means and variances or the distributions of the random variables are assumed to be known *a priori*. The ultimate goal of the stochastic analysis is then to find the sequence that has the "best" statistical distribution. Knowing such a distribution will enable the management to plan for capacity and quote delivery dates in a manner that achieves set target service levels and higher customer satisfaction. However, finding the distribution of a scheduling criterion is extremely complex and, at times, practically impossible. Hence, researchers resort to more modest and practically viable criteria. The objective could then be to optimize some function of the performance measure of interest. The performance measure is also a random variable because it is a function of the input variables, which are given to be random. Predominantly, this function of the output performance measure is its expectation: that is, the goal is to optimize (minimize or maximize) the expectation of the performance measure.

The reason for such an approach can be surmised easily from the fact that computing or formulating the expectation function is relatively easier and less complex than computing or formulating any other function of the random variable, for example, its variance. In addition, optimization becomes arduous and may even become impossible with the incorporation of the variance function. Furthermore, determining the variance of a performance measure is highly complicated and laborious and is not straightforward for most of the commonly used performance measures (e.g., makespan, tardiness,) in scheduling. Hence, a preponderance of the work in stochastic scheduling has dealt with optimizing the expected value of a performance measure. To cite a simple example, while scheduling jobs with random processing times on a single machine with completion time as the performance measure, the predominant motive is to minimize the total expected completion time of all the jobs. By focusing only on the expected value and ignoring the variance of the objective, the scheduling problem becomes purely deterministic, and the significant ramifications of schedule variability are neglected. However, in many practical cases, a scheduler would prefer to have a stable schedule with minimum variance over a schedule that has lower expected value and unknown (and possibly higher) variance.

1.4 Significance of Variance in Scheduling

As mentioned earlier, it is important to consider the issue of the variance of a performance measure in scheduling problems. To illustrate the significance of variance by means of a very simple example, consider four jobs waiting to be processed on a single machine with the objective of minimizing the total completion time (total flow time). The job processing times (in some specified units) are random with known means and variances, as given in Table 1.1.

Conventionally, the approach in tackling this problem would be to minimize the total expected completion time by sequencing the jobs using the shortest expected processing time (SEPT) policy. Hence the optimal SEPT sequence is 3-4-1-2. If we let C_j be the completion time of job j, then the resulting expectation and variance of the total completion time are $E\left[\sum C_j\right] = 256$ and $\mathrm{Var}\left[\sum C_j\right] = 357$.

However, by scheduling the jobs alternatively, say, in the 4-3-1-2 sequence, we have $E\left[\sum C_j\right] = 258$ and $\mathrm{Var}\left[\sum C_j\right] = 217$. Schedule 2 possesses a slightly higher expected value but has a considerably lower variance than the SEPT schedule. This is illustrated in Figure 1.1, assuming that $\sum C_j$ for the two schedules follow a normal distribution.

If the manufacturer prefers to deliver all the jobs by a particular date, say, $d = 280$, we then can analyze the probability with which the deadline will be met using the two schedules in Figure 1.1:

Schedule 1 (SEPT schedule): $\mu_1 = 256$ and $\sigma_1^2 = 357$

$$\mathrm{Pr}\left[\sum C_j \leq 280\right] = \mathrm{Pr}[Z \leq 1.272] = 0.898 = 89.8\%$$

Schedule 2: $\mu_2 = 258$ and $\sigma_2^2 = 217$

$$\mathrm{Pr}\left[\sum C_j \leq 280\right] = \mathrm{Pr}[Z \leq 1.493] = 0.9324 = 93.24\%$$

(Z is the standard normal variable with mean 0 and variance 1.)

Table 1.1. *Total Completion Time Example*

N	1	2	3	4
μ	35	40	20	22
σ^2	8	5	20	0

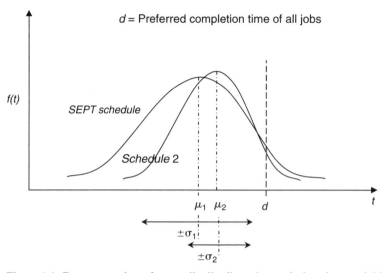

Figure 1.1. Representation of normally distributed completion time variables.

The probability of meeting the deadline is higher if the second schedule is employed. Apparently, it becomes imperative to determine a sequence that is "good" in terms of both expectation and variance. We would not have been able to identify Schedule 2, which, in fact, turned out to be practically better, had we not included the variance and, instead, had focused only on the expectation.

Soroush and Fredenhall (1984) recognized the importance of considering both mean and variance in scheduling while studying the impact of random processing times on the earliness and tardiness costs for scheduling jobs on a single machine. The significance of variance is not necessarily confined to production systems because it has been addressed in the context of other fields as well, such as telecommunication networks (Shayman and Gaucherand, 2001), financial investment decision problems (Chue and Nagasawa, 1999), and media planning/selection (de Kluyver, 1980). In communication networks, *sequential testing* is the process of identifying the defective component from a set of components that is attributed as a root cause of a failure. There is a random cost associated with the testing of each component, and traditionally, the objective has been to find a sequence that minimizes the average (or expected) sum of testing costs. Shayman and Gaucherand (2001) assert that the network scheduler should use a risk-sensitive optimality criterion to correctly model the system by taking into consideration the risk factor associated with the variance of the total cost. In optimal investment decision problems, the most desirable objective for a decision maker is to maximize the expected profit resulting from the investment as well as minimize the investment risk (variation in profit). In media scheduling problems, the objective is to select and schedule different media options that would maximize the return (e.g., gross profit, gross audience) given a set of media options, budget, and other relevant data. In addition, it is necessary to recognize the effect of variance and control schedule variance by incorporating an effective risk-return analysis (de Kluyver, 1980; de Kluyver and Baird, 1984).

1.5 Multiobjective or Multicriteria Stochastic Scheduling

Multiobjective or multicriteria optimization, especially in the field of scheduling, has always been an interesting and challenging topic for researchers. A scheduler's endeavor, from a practical point of view, is to optimize one or more objectives of interest simultaneously and achieve a trade-off solution, which is commonly referred to as a *Pareto-optimal solution*. The solution to a multiobjective optimization problem is considered to be Pareto-optimal if there are no other solutions that are better in satisfying all the objectives simultaneously. That is, there can be other solutions that are better in satisfying one or several objectives, but they must be worse than the Pareto-optimal solution in satisfying the remaining objectives.

Deterministic multiobjective scheduling has been addressed rather extensively in the literature, and some of the work reported can be found in Wassenhove and Gelders (1980), Lin (1983), Nelson et al. (1986), Daniels and Chambers (1990), Sarin and Hariharan (2000), and Sarin and Prakash (2004), among many others. T'Kindt and Billaut (2005) have recently edited a special issue of the *European Journal of Operational Research* devoted to this topic.

On the stochastic front, Forst (1995) addressed the problem of minimizing the sum of the expected total weighted tardiness and the expected total weighted flow time for the single-machine and m-machine flow-shop scheduling problems. He proved that an optimal sequence is obtained by sequencing the jobs in increasing stochastic order of their processing times. The job processing times are assumed to be independent random variables, and the jobs have a common random due date. Lin and Lee (1995) considered a single-machine scheduling problem with known distributions of random processing times and due dates. The objective was to determine a schedule that minimized a secondary criterion subject to a primary criterion that was held at its best value. They formulated three different models with completion times and lateness-related bicriteria objectives, and they provided algorithms for obtaining optimal solutions.

Few studies have been devoted to the stochastic analysis of a schedule as compared with its deterministic counterpart. Liu et al. (1992) dealt with a discounted Markov decision model to determine a schedule with optimal expectation and variance of a criterion. They discussed the difficulties involved in minimizing variance by Markovian models. They also formulated a multiobjective nonlinear programming problem and presented an algorithm for determining a Pareto-optimal solution.

Lu et al. (1994) addressed the problem of reducing the mean and variance of cycle times in semiconductor manufacturing environments, which feature the characteristic reentrant process flows. In reentrant flows, lots repeatedly return to the same service stations for further processing at different stages of their production. Lu et al. introduce a new class of scheduling policies, called *fluctuation smoothing policies*, that achieve the best mean and variance of the cycle time. The effectiveness of these policies was demonstrated via simulation modeling of two semiconductor manufacturing plants. Kumar and Kumar (1994), subsequently, established through their work that these policies are stable for all stochastic reentrant lines under certain conditions.

It would be germane at this juncture to mull over the fact that the multiple objectives that the researchers considered in stochastic multicriteria scheduling are related predominantly to completion time and tardiness. From a problem-modeling perspective, the different scheduling parameters, *viz.,* job processing times, due dates, and so on, were primarily modeled as random variables with known distributions. The performance measures of interest, such as the total flow time or tardiness, which are in turn random variables and a function of the random variables, were optimized. More often than not, this function is the expectation. This seemed valid enough because consideration of other functions, such as the variance of completion time or lateness, would make the problem enormously complex and difficult to solve.

1.6 Variance of the Performance Measure: Other Production Systems

As we strive to understand the importance of variance from a scheduling perspective, it is also apposite to survey the variance-related research in other production control systems. This section briefly reviews the work done in computing the variance of the output measure in serial production lines operated using CONWIP and other systems. There is an abundance of information in this domain, and only an illustrative review is provided to underline the fact that variance is indeed a parameter worthy of consideration.

1.6.1 CONWIP Systems

A CONWIP line, or *con*stant *work-in-p*rocess line, is a pull-based production system proposed by Spearman et al. (1990) (Figure 1.2). The output measures in a CONWIP line are, predominantly, throughput [or time between departures (TBD)] and flow time. Considerable research has been done on CONWIP systems to study and analyze the mean and variance of these measures. Spearman

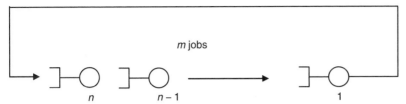

Figure 1.2. CONWIP schematic.

and Hopp (1991) developed an expression to estimate the throughput of a CONWIP manufacturing line subject to machine failures. They computed the throughput and average cycle time as a function of the work-in-progress (WIP) level.

In a subsequent paper, Duenyas et al. (1993) derived an approximation for the variance of the throughput of a CONWIP line with deterministic processing and random outages. Dar-el et al. (1998) focused on a CONWIP line to develop estimates for four important performance measures: the means and variances of the TBD and flow time. TBD is the inverse of the throughput rate.

1.6.2 Production Lines

It has been shown that the distribution of the output from a production line is asymptotically normal as a result of the central limit theorem. Hence, a majority of the work dealing with uncertain production lines, owing to random processing times or unreliable machines, had striven only to determine the expectation and variance of the output performance measure of interest. Knowing the mean and variance of the output gives the asymptotic distribution of the throughput, which could be used to derive other performance measures (e.g., meeting a customer due date) based on the probability of other events.

Hendricks (1992) analyzed the mean and variance of the output process of a serial production line of N machines with exponential processing time distributions and finite buffer capacities. Analytic expressions for the interdeparture distribution and the correlation structure of the output process were developed using a continuous-time Markov chain model.

Tan (1997) developed a closed-form expression for the variance of the throughput of an N-station production line with no intermediate buffers and time-dependent failures. Time-to-failure and time-to-repair distributions were assumed to be exponential, and the variance of the throughput was determined by modeling the process as an irreducible recurrent Markov process. In a subsequent paper, Tan (2000) determined the variance of the throughput of a production line with finite buffers by modeling the line as a discrete-time Markov chain.

1.7 Processing Time Variance in Scheduling

Variation in the processing times is a major factor or cause of uncertainty in scheduling, and the impact of variation in processing times on the efficiency of scheduling has been a subject of discussion in the literature for a long time. McKay et al. (1988) pointed out that the primary reason for poor applicability of scheduling theory in practice is its inability to properly account for extreme variations in processing times. This is primarily due to the ubiquitous attempt to use deterministic models in practical situations, which are highly stochastic. In the world of agile and lean manufacturing, effective scheduling under uncertainty has become a survival necessity for companies to meet committed shipping dates and effective utilization of available resources. Hence, it becomes imperative to devise the right scheduling strategies to employ in practice.

Dodin (1996) contends that the pseudodeterministic sequence that is obtained by sequencing tasks when all the activities are assumed to take their expected times does not fully reflect the goals of stochastic analysis of a schedule. Dodin further suggests using an alternative sequence determined using a ranking system based on *optimality indices* (OIs), defined as their respective probabilities of being the best. The OIs are computed using the strong dominance properties of distributions. Dodin conducted extensive simulations to analyze and compare the sequences obtained by the preceding two methods in order to determine which performs best. However, the results do not favor one method over the other and remain inconclusive. Portuogal and Trietsh (1998) also agreed with Dodin in stating that minimizing the expectation alone is not good enough for the scheduler. They introduced and defined two new sequences called *stochastically smallest* and *almost surely smallest sequences*. They analyzed these sequences along with Dodin's two sequences, under different scenarios, to find the sequence with the best distribution. They concluded that stochastically smallest and almost surely smallest should always be selected whenever they exist. However, stochastically smallest and almost surely smallest sequences do not always exist in practice, and even if they do exist, determining them is an arduous task. They argued that Dodin's sequence based on OIs does not take variability into account and suggested that a variance-reduction objective should be considered explicitly to attain optimal service levels while retaining the expected completion time.

Ayhan and Olsen (2000) considered scheduling on a single machine (server) that processes a number of different classes of items. They proposed two heuristic procedures for scheduling on such a multiclass single server that minimize the throughput time variance and the outer percentiles of the throughput time.

1.8 Analytic Evaluation of Expectation and Variance of a Performance Measure

In addition to these observations, a telling inference that can be made is that analytic expressions for the expectation and variance of simpler performance measures such as the total completion time on a single machine can be readily formulated and computed, but expressions for other complex measures related to tardiness are relatively complex and not as straightforward to evaluate. This task is even harder for measures related to makespan in multimachine scheduling environments such as parallel machines, flow shops, or job shops. Besides, no comprehensive work is reported in the literature that strives to address this issue. Hence, our primary motive in this book is to present a comprehensive analysis in order to devise methodologies and derive closed-form expressions (wherever possible) to determine the expectation and variance of various performance measures for different scheduling environments. The scheduling environments considered in our analysis include

1. Scheduling on a single machine
2. Permutation flow shops with unlimited intermediate storage
3. Job shops with unlimited intermediate storage
4. Scheduling on identical machines in parallel.

This analysis is contingent on the facts that the schedule is given *a priori* and that it is necessary to ascertain the expectation and variance of the given performance measure for that given schedule. The position of each job is, therefore, known with certainty from the schedule. The randomness in the scheduling process is due to variable processing times with known means and variances. All other parameters, such as the job due dates and weights, are assumed to be deterministic. In some cases, it might also be necessary to know the processing time distributions, and those instances are cited appropriately. Our interest, then, is to develop analytic expressions or methodologies to compute the parameters under consideration, *viz.,* expectation and variance of the objective function value. The analysis does not involve optimization, and it is a vital exercise in modeling the performance of the scheduling system. However, it is worth mentioning that this endeavor will only trigger and enable variance considerations in schedule optimization. This knowledge would provide valuable insights in improving the performance of a schedule. A scheduler would be in a better position to base his or her decisions knowing the variability of the schedule and appropriately striking a balance between the expected value and variance. In addition, these expressions and methodologies can be incorporated in various scheduling algorithms (and available software packages) to determine efficient schedules in terms of both the expectation and variance.

The different models considered for our analysis include

1. **Single-machine models.** The different performance measures considered for the single-machine case can be classified under two different categories, namely,
   a. Completion-time-based measures
   b. Tardiness-based measures.
   The various completion-time-based measures are
   a. Total completion time (total flow time)
   b. Total weighted completion time
   c. Total weighted discounted completion time.
   The various tardiness-based measures are
   a. Total tardiness
   b. Total weighted tardiness
   c. Total number of tardy jobs
   d. Total weighted number of tardy jobs
   e. Mean lateness
   f. Maximum lateness.
2. **Parallel-machine models.** For parallel machines, both preemption and no-preemption cases are considered. The performance measures are makespan and total completion time for the no-preemption case, and only makespan for the preemption case.
3. **Flow shops.** The objective is the makespan of a permutation flow shop with unlimited intermediate storage.
4. **Job shops.** The objective is to evaluate the makespan for a classic job shop with unlimited intermediate storage.

However, before we present our work, we would like to study and understand in detail the available literature in the field of stochastic scheduling, where researchers have attempted to consider both expectation and variance in their analyses. Our focus is only on the randomness owing to processing time uncertainty, which, as mentioned earlier, is a significant cause of stochasticity in scheduling.

1.9 Organization of the Book

The organization of this book is as follows: In this introductory chapter we have focused on the impact of uncertainty in scheduling and the need for efficient modeling of stochastic scheduling problems, as well as the need to devise effective scheduling strategies to counter the impact of uncertainty (or variability). We have specifically highlighted the prevalence of variability in job processing times and elicited the issue of neglecting variance in schedule optimization and the significance of considering variance. The need for a comprehensive analytic evaluation of the expectation and variance of the different performance measures has also been enunciated. Chapter 2 deals with robust scheduling approaches to hedge against processing time uncertainty. Various model formulations and solution methodologies available in the literature are presented in detail. Chapter 3 deals with the scheduling techniques available for determining a set of "nondominated schedules" or "expectation-variance efficient sequences" and the methodologies used to select a preferred schedule from such a set. We discuss our work in Chapters 4 through 8, wherein a comprehensive analysis of schedules for various scheduling environments and performance measures is presented. Chapters 4 through 7, respectively, deal with scheduling in a single-machine, flow-shops, job-shops, and parallel-machine environments. We develop closed-form expressions (wherever possible) and devise methodologies to evaluate the expectation and variance of various performance measures. The methodologies are also illustrated using example problems. The closed-form expressions for many performance measures rely on the assumption of normal distributions for job processing times. We relax this assumption in Chapter 8 and consider the case of general processing time distributions. Our analysis for this case is based on the use of finite mixture models. Concluding remarks are then made in Chapter 9.

We have also developed a software package to help implement the methodologies developed in Chapters 4 through 8. This is a user-friendly software program termed *XVA-sched* (for the *ex*pectation-*v*ariance *a*nalysis of a *sched*ule). This software is contained on the CD that accompanies this book. The instructions to use this software are included in the Appendix (Section A.4).

2 Robust Scheduling Approaches to Hedge against Processing Time Uncertainty

2.1 Introduction

Robust scheduling is one of the approaches adopted to deal with uncertainties in stochastic scheduling. Since these uncertainties, arising as they do from variability in the scheduling parameters, significantly affect the performance of a system, a viable approach has been to determine a robust schedule that is least susceptible to disturbances or variations in the input parameters and that has minimum variability in its output performance measure. For instance, when processing times are uncertain, a robust schedule would have the least variance of the output performance measure, although it might not necessarily have the least expected value. As also mentioned in Chapter 1, of the various parameters involved in scheduling, we intend to focus on the uncertainty factor associated with job processing times. The following sections detail different robust scheduling methodologies that are available to model certain scheduling environments and the underlying principles behind those various approaches. Our primary focus is to elicit and delve in detail on the modeling approaches and not to expostulate the solution methodologies. However, we do briefly discuss the various solution approaches that are used to solve these robust scheduling problems and also state relevant results (as propositions) that are used in developing these solution methodologies. We intend to capture the core techniques that have been used in modeling robustness in scheduling and would advise interested readers to refer to the appropriate literature for a detailed and complete exposition.

2.2 Modeling Processing Time Uncertainty

There are a number of ways to model processing time uncertainty in stochastic scheduling. Predominantly, the job processing times are modeled as random variables with the cognizance of their entire distribution or just their means and variances. Furthermore, processing time variability also can be captured in the following ways:

1. Discrete processing time scenarios
2. Continuous processing time intervals

In the case of *discrete processing time scenarios,* there is a set of possible scenarios Λ. For a scenario $\lambda \in \Lambda$, each job i realizes a specific processing time denoted by p_i^λ. P^λ is a vector representing the processing times of all the jobs corresponding to scenario λ. On the other hand, in the case of *continuous processing time intervals,* the variability in processing times is captured by specifying a range $[\underline{p_i}, \overline{p_i}]$ of realizable times, where $\underline{p_i}$ and $\overline{p_i}$, respectively, represent the lower and upper bounds on the processing time of job i. An infinite set of processing time scenarios can be derived in this case. These two approaches have been used commonly to develop robust scheduling models, and we will examine those models in detail in the following sections.

2.3 Robust Scheduling for Single-Machine Systems

Scheduling on a single machine is the simplest and easiest of all scheduling problems. The problem is to schedule n jobs on a single machine for the objective of optimizing a performance criterion of interest. We first consider the robust scheduling single-machine model with discrete processing time scenarios. As stated earlier, the n independent jobs to be scheduled realize a specific processing time p_i^λ in each scenario, where i and λ are the indices for the jobs and scenarios, respectively. Given the performance measure of total flow time, the optimal schedule λ for a given scenario can be found easily by arranging the jobs in nondecreasing order of their processing times to result in the shortest processing time (SPT) schedule. Let G^λ represent the value of the total flow time corresponding to this optimal schedule. The notion of schedule robustness or the schedule that has the best worst-case performance is based on determining a schedule X that minimizes the worst-case deviation from optimality for the performance criterion of interest $(\varphi(X, P^\lambda))$ with respect to all possible scenarios Λ.

An absolute deviation robust scheduling problem (ADRSP) to determine a robust schedule X can be formulated as follows (Daniels and Kouvelis, 1995):

$$\text{ADRSP: } \min_{X}\{\max_{\lambda \in \Lambda} |\varphi(X, P^\lambda) - G^\lambda|\}$$

subject to

$$\sum_{k=1}^{n} x_{ik} = 1, \quad i = 1, 2, \ldots, n \tag{2.1}$$

$$\sum_{i=1}^{n} x_{ik} = 1, \quad k = 1, 2, \ldots, n \tag{2.2}$$

$$x_{ik} \in \{0, 1\}, \quad i = 1, 2, \ldots, n \quad k = 1, 2, \ldots, n \tag{2.3}$$

where $x_{ik} = \begin{cases} 1, & \text{if job } i \text{ is in the } k\text{th position} \\ 0, & \text{otherwise} \end{cases}$

The performance measure, the total flow time, is given by

$$\varphi(X, P^\lambda) = \sum_{i=1}^{n} \sum_{k=1}^{n} (n - k + 1) p_i^\lambda x_{ik}$$

Constraints (2.1) and (2.2) capture the fact that each job occupies only one position and each position contains only one job, respectively. A minor variation of the preceding problem, called the *relative deviation robust scheduling problem* (RDRSP), is to determine a schedule that minimizes the worst-case percentage deviation from optimality for the total flow time performance criterion. The only deviation from the preceding formulation is that the difference $(\varphi(X, P^\lambda) - G^\lambda)$ in the preceding objective function is replaced by the ratio $\varphi(X, P^\lambda)/G^\lambda$.

An equivalent form of the ADRSP is

ADRSP′: min z

subject to

$$\sum_{i=1}^{n} \sum_{k=1}^{n} (n - k + 1) p_i^\lambda x_{ik} \leq z + G^\lambda, \quad \forall \lambda \in \Lambda$$

and constraints (2.1), (2.2), and (2.3).

The robust scheduling models ADRSP and ADRSP′ can be shown to be NP-hard (Daniels and Kouvelis, 1995). This essentially follows from the equivalence of the ADRSP model with the assignment problem with a single side constraint, a problem that has been shown to be NP-hard (see Mazzola and Nebbe, 1986). Since the problems assume discrete processing time scenarios, it becomes cumbersome to evaluate all possible scenarios when $|\Lambda|$ is large. Hence it might be preferable to use the continuous processing time interval case and specify ranges for the independent jobs. Modeling the problem using this method can help to unearth certain interesting properties that are useful in the solution procedures. The next section briefly discusses these properties of robust schedules, as given in Daniels and Kouvelis (1995) and presented here as Propositions 2.1, 2.2, and 2.3.

2.3.1 Properties of Robust Schedules

2.3.1.1 Worst-Case Absolute Deviation

Under the continuous processing time interval case, the jobs have an upper bound and a lower bound on their processing times. A processing time scenario λ is defined as an extreme point scenario if and only if for all jobs $i = 1, 2, \ldots, n$, either $p_i^\lambda = \underline{p_i}$ or $p_i^\lambda = \overline{p_i}$. Let ξ denote a sequence $\{\xi_{[1]}, \xi_{[2]}, \ldots, \xi_{[n]}\}$, where $\xi_{[k]}$ represents the job that occupies position k in sequence ξ. Also, let $\pi(i)$ represent the position occupied by job i in sequence ξ and $\pi_\lambda^*(i)$ be the position occupied by the same job in the SPT sequence for processing time scenario λ.

Proposition 2.1. For any sequence ξ and the total flow time performance criterion, the scenario that maximizes the absolute deviation from the optimal value belongs to the set of extreme point scenarios.

Proposition 2.2. Given λ_0 to be the worst-case absolute deviation scenario for the sequence ξ, the following are true: $p_i^{\lambda_0} = \overline{p_i}$, when $\pi_{\lambda_0}^*(i) > \pi(i)$, and $p_i^{\lambda_0} = \underline{p_i}$, when $\pi_{\lambda_0}^*(i) \leq \pi(i)$.

These two propositions can be used to formulate a problem called the *worst-case absolute deviation problem* (WCADP) for a given sequence ξ.

$$\text{Let } y_{ik} = \begin{cases} 1, & \text{if } \pi_{\lambda_0}^*(\xi_{[j]}) = k \\ 0, & \text{otherwise, for } j = 1, 2, \ldots, n \text{ and } k = 1, 2, \ldots, n \end{cases}$$

Then the WCADP formulation is given by

$$\text{WCADP: } \max \sum_{j=1}^{n} \left\{ \sum_{k=1}^{j} [k-j] \underline{p}_{\xi_{[j]}} y_{jk} + \sum_{k=j}^{n} [k-j] \overline{p}_{\xi_{[j]}} y_{jk} \right\} \tag{2.4}$$

subject to

$$\sum_{k=1}^{n} y_{jk} = 1, \quad j = 1, 2, \ldots, n \tag{2.5}$$

$$\sum_{j=1}^{n} y_{jk} = 1, \quad k = 1, 2, \ldots, n \tag{2.6}$$

$$y_{jk} \in \{0, 1\}, \quad j = 1, 2, \ldots, n \quad k = 1, 2, \ldots, n \tag{2.7}$$

In contrast to the ADRSP, the pure assignment problem structure of the WCADP makes it solvable in polynomial time. The significance of the formulation of the WCADP is in determining bounds for use in a branch-and-bound solution procedure that can be developed to solve the ADRSP. We present this procedure in Section 2.3.2 below.

2.3.1.2 Other Properties of Robust Schedules

The following proposition demonstrates that the relative positions of jobs i and j in the robust schedule can be unambiguously determined if the processing time intervals of two jobs either do not overlap or overlap in part.

Proposition 2.3. If for two jobs i and j, $\underline{p_i} \leq \underline{p_j}$ and $\overline{p_i} \leq \overline{p_j}$, then there exists an absolute robust single-machine sequence for the total flow time performance criterion in which job i precedes job j.

2.3.2 Solution Approaches for ADRSP

2.3.2.1 Branch-and-Bound Algorithm for ADRSP

A branch-and-bound procedure based on the various properties of robust schedules presented above and a surrogate relaxation approach (see Glover, 1975) can be used to solve the single-machine robust scheduling problem.

The surrogate relaxation of the absolute deviation robust scheduling problem (ADRSP') is given as

$$SA(\gamma): \min z$$

subject to

$$\sum_{\lambda \in \Lambda} \gamma_\lambda \sum_{i=1}^{n} \sum_{k=1}^{n} (n-k+1) p_i^\lambda x_{ik} \leq z + \sum_{\lambda \in \Lambda} \gamma_\lambda G_\lambda$$

and constraints (2.1), (2.2), and (2.3).

Where $\gamma = \{\gamma_\lambda : \lambda \in \Lambda\}$ is a vector of multipliers such that $\gamma_\lambda \geq 0$ and $\sum_{\lambda \in \Lambda} \gamma_\lambda = 1$.

The following equivalent problem, $SAE(\gamma)$, can be solved to determine the optimal solution to $SA(\gamma)$.

$$SAE(\gamma): \min \sum_{i=1}^{n} \sum_{k=1}^{n} (n-k+1) \left(\sum_{\lambda \in \Lambda} \gamma_\lambda p_i^\lambda \right) x_{ik}$$

subject to

constraints (2.1), (2.2), and (2.3).

If z_a^* denotes the optimal objective value for the ADRSP', then we can write

$$SA(\gamma) = SAE(\gamma) - \sum_{\lambda \in \Lambda} \gamma_\lambda G_\lambda \leq z_a^*.$$

A surrogate bounding approach based on the solutions to the $SAE(\gamma)$ and the WCADPs formulated earlier can provide lower and upper bounds to the

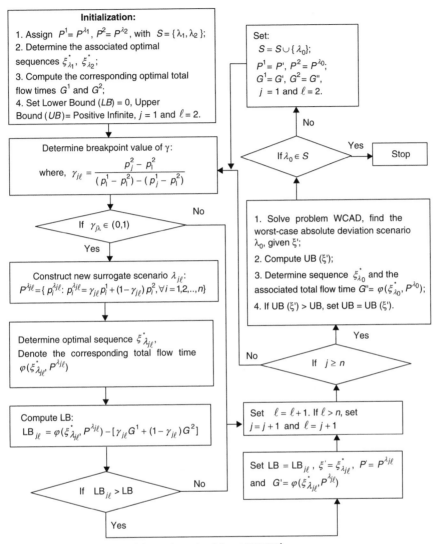

Figure 2.1. A branch-and-bound method for ADRSP'.

ADRSP'. This bounding scheme in conjunction with the dominant properties (Propositions 2.1, 2.2, and 2.3) can be incorporated in a branch-and-bound method to solve the original ADRSP'. This is shown in Figure 2.1.

2.3.2.2 Heuristic Approaches for ADRSP

A couple of heuristic approaches designated as the *endpoint sum* and *endpoint product* also can be used to obtain approximate solutions to the ADRSP' (see Daniels and Kouvelis, 1995). The *endpoint sum* heuristic computes the sum of the lower and upper bounds on the processing times for each job and sorts the set in nondecreasing order of this sum to obtain a sequence ξ, whereas the *endpoint product* heuristic, on the other hand, computes the product of the bounds on the processing times of the jobs and sorts them similarly. The associated worst-case absolute deviation from optimality for each of these sequences can be determined by solving the WCADP for the corresponding sequence.

2.3.2.3 Dynamic Programming Approach to Solve the Robust Scheduling Problem with a Finite Number of Scenarios

When the number of processing time scenarios is finite, the single-machine robust scheduling problem is NP-complete and can be solved exactly using a dynamic programming approach (see Yang and Yu, 2002). A brief outline of the dynamic programming approach is as follows:

Let $f^\lambda(\xi)$ denote the objective value (the total flow time) of any given sequence ξ corresponding to scenario λ. The state variables constitute a 0–1 vector $u = (u_1, u_2, \ldots, u_n)$ that captures whether or not each job i has been put in the processing time sequence and a vector $\alpha = (\alpha^1, \alpha^2, \ldots, \alpha^{|A|})$ that contains the contribution of the jobs already in the sequence to the total flow times under all scenarios. Two indexed sets are defined for a vector u : $I(u) = \{i | u_i = 1,\ i \in \{1, 2, \ldots, n\}\}$, and $O(u) = \{i | u_i = 0,\ i \in \{1, 2, \ldots, n\}\}$. For $i \in O(u)$, a new vector, $u_{(+i)}$, is defined that satisfies $I(u) \cup \{i\}$.

Define $f(u, \alpha)$ to be the minimum of the $\max_{\lambda \in A} f^\lambda(\gamma)$ value of the sequence among all sequences whose first $|I(u)|$ jobs are the job i's in the set $I(u)$, and note that these jobs' contribution to the total completion time in scenario λ is $\alpha^\lambda,\ \forall \lambda \in A$.

The initial condition is $f(1_n, \alpha) = \max_{\lambda \in A} \alpha^\lambda$

The recursive equation for the dynamic programming procedure is as follows:

$$f(u, \alpha) = \min_{i \in O(u)} f\left(u_{(+i)};\ \alpha^1 + |O(u)|p_i^1, \ldots, \alpha^{|A|} + |O(u)|p_i^{|A|}\right)$$

where $|O(u)|$ is the cardinality of $O(u)$. At stage k, we will find at least one $i \in O(u_k)$ that satisfies

$$f(u_k, \alpha_k) = f\left(u_{k(+i)};\ \alpha_k^1 + (n-k)p_i^1, \ldots, \alpha_k^{|A|}\right) + \left((n-k)p_i^{|A|}\right)$$

The recursive procedure then is updated as follows:

$$u_{k+1} = u_{k(+i)}, \quad \alpha_{k+1}^\lambda = \alpha_k^\lambda + (n-k)p_i^\lambda, \quad \forall \lambda \in A$$
$$k = k + 1, \quad \sigma(k) = i$$

where $\sigma(k)$ is the job in the kth position of the optimal sequence. The procedure continues until $k = n$. The optimal value for the absolute deviation problem is given by $f(0_n, 0^{|A|})$.

To illustrate this procedure, consider the following example consisting of three jobs and five scenarios (Table 2.1).

Table 2.1. *Processing Time Data for an ADRSP*

	Scenario 1	Scenario 2	Scenario 3	Scenario 4	Scenario 5
Job 1	4	6	7	4	5
Job 2	2	6	5	5	4
Job 3	7	7	6	6	9

Stage 0:

We have $u_0 = 0_n$ and $\alpha_0^\lambda = 0, \forall \lambda \in \Lambda$. Then

$$f(u_0, \alpha_0) = f(0_n, 0^{|\Lambda|})$$

$$= \min_{i \in O(u)} f\left(u_{(+i)}; \alpha^1 + |O(u)|p_i^1, \ldots, \alpha^{|\Lambda|} + |O(u)|p_i^{|\Lambda|}\right)$$

$$= \min\{f((1,0,0); 0+3 \times 4, 0+3 \times 6, 0+3 \times 7, 0+3 \times 4, 0+3 \times 5)$$

$$+ f((0,1,0); 0+3 \times 2, 0+3 \times 6, 0+3 \times 5, 0+3 \times 5, 0+3 \times 4)$$

$$+ f((0,0,1); 0+3 \times 7, 0+3 \times 7, 0+3 \times 6, 0+3 \times 6, 0+3 \times 9)\}$$

$$= \min\{f((1,0,0); 12, 18, 21, 12, 15) + f((0,1,0); 6, 18, 15, 15, 12)$$

$$+ f((0,0,1); 21, 21, 18, 18, 27)\}$$

Stage 1:

If $u_1 = (1,0,0)$ and $\alpha_1 = (12, 18, 21, 12, 15)$, we have $O(u_1) = \{2, 3\}$. Then

$$f(u_1, \alpha_1) = f((1,0,0); 12, 18, 21, 12, 15)$$

$$= \min\{f((1,1,0); 12+2 \times 2, 18+2 \times 6, 21+2 \times 5, 12+2 \times 5,$$

$$15+2 \times 4) + f((1,0,1); 12+2 \times 7, 18+2 \times 7, 21+2 \times 6,$$

$$12+2 \times 6, 15+2 \times 9)\}$$

$$= \min\{f((1,1,0); 16, 30, 31, 22, 23) + f((1,0,1); 26, 32, 33, 24, 33)\}$$

If $u_1 = (0,1,0)$ and $\alpha_1 = (6, 18, 15, 15, 12)$, we have $O(u_1) = \{1, 3\}$. Then

$$f(u_1, \alpha_1) = f((0,1,0); 6, 18, 15, 15, 12)$$

$$= \min\{f((1,1,0); 6+2 \times 4, 18+2 \times 6, 15+2 \times 7, 15+2 \times 4,$$

$$12+2 \times 5) + f((0,1,1); 6+2 \times 7, 18+2 \times 7, 15+2 \times 6,$$

$$15+2 \times 6, 12+2 \times 9)\}$$

$$= \min\{f((1,1,0); 14, 30, 29, 23, 22) + f((0,1,1); 20, 32, 27, 27, 30)\}$$

If $u_1 = (0,0,1)$ and $\alpha_1 = (21,21,18,18,27)$, we have $O(u_1) = \{1,2\}$. Then

$$
\begin{aligned}
f(u_1,\alpha_1) &= f((0,0,1);21,21,18,18,27)\\
&= \min\{f((1,0,1);\ 21+2\times4,21+2\times6,18+2\times7,18+2\times4,\\
&\qquad 27+2\times5)+f((0,1,1);\ 21+2\times2,21+2\times6,18+2\times5,\\
&\qquad 18+2\times5,27+2\times4)\}\\
&= \min\{f((1,0,1);29,33,32,26,37)+f((0,1,1);25,33,28,28,35)\}
\end{aligned}
$$

Stage 2:
If $u_2 = (1,1,0)$ and $\alpha_2 = (16,30,31,22,23)$, we have $O(u_2) = \{3\}$. Then

$$
\begin{aligned}
f(u_2,\alpha_2) &= f((1,1,0);16,30,31,22,23)\\
&= \min\{f((1,1,1);\ 16+1\times7,30+1\times7,31+1\times6,22+1\times6,\\
&\qquad 23+1\times9)\}\\
&= f((1,1,1);\ 23,37,37,28,32)\\
&= \max\{23,37,37,28,32\}\ \text{(Stage 3)}\\
&= 37
\end{aligned}
$$

If $u_2 = (1,1,0)$ and $\alpha_2 = (14,30,29,23,22)$, we have $O(u_2) = \{3\}$. Then

$$
\begin{aligned}
f(u_2,\alpha_2) &= f((1,1,0);14,30,29,23,22)\\
&= \min\{f((1,1,1);\ 14+1\times7,30+1\times7,29+1\times6,23+1\times6,\\
&\qquad 22+1\times9)\}\\
&= f((1,1,1);\ 21,37,35,29,31)\\
&= \max\{21,37,35,29,31\}\ \text{(Stage 3)}\\
&= 37
\end{aligned}
$$

If $u_2 = (1,0,1)$ and $\alpha_2 = (26,32,33,24,33)$, we have $O(u_2) = \{2\}$. Then

$$
\begin{aligned}
f(u_2,\alpha_2) &= f((1,0,1);26,32,33,24,33)\\
&= \min\{f((1,1,1);\ 26+1\times2,32+1\times6,33+1\times5,24+1\times5,\\
&\qquad 33+1\times4)\}\\
&= f((1,1,1);\ 28,38,38,29,37)\\
&= \max\{28,38,38,29,37\}\ \text{(Stage 3)}\\
&= 38
\end{aligned}
$$

If $u_2 = (1,0,1)$ and $\alpha_2 = (29,33,32,26,37)$, we have $O(u_2) = \{2\}$. Then

$$f(u_2,\alpha_2) = f((1,0,1);29,33,32,26,37)$$
$$= \min\{f((1,1,1);\ 29+1\times 2,33+1\times 6,32+1\times 5,26+1\times 5,$$
$$37+1\times 4)\}$$
$$= f((1,1,1);\ 31,39,37,31,41)$$
$$= \max\{31,39,37,31,41\} \text{ (Stage 3)}$$
$$= 41$$

If $u_2 = (0,1,1)$ and $\alpha_2 = (20,32,27,27,30)$, we have $O(u_2) = \{1\}$. Then

$$f(u_2,\alpha_2) = f((0,1,1);20,32,27,27,30)$$
$$= \min\{f((1,1,1);\ 20+1\times 4,32+1\times 6,27+1\times 7,27+1\times 4,$$
$$30+1\times 5)\}$$
$$= f((1,1,1);\ 24,38,34,31,35)$$
$$= \max\{24,38,34,31,35\} \text{ (Stage 3)}$$
$$= 38$$

If $u_2 = (0,1,1)$ and $\alpha_2 = (25,33,28,28,35)$, we have $O(u_2) = \{1\}$. Then

$$f(u_2,\alpha_2) = f((0,1,1);25,33,28,28,35)$$
$$= \min\{f((1,1,1);\ 25+1\times 4,33+1\times 6,28+1\times 7,28+1\times 4,$$
$$35+1\times 5)\}$$
$$= f((1,1,1);\ 29,39,35,32,40)$$
$$= \max\{29,39,35,32,40\} \text{ (Stage 3)}$$
$$= 40$$

Tracking back to Stage 1:
If $u_1 = (1,0,0)$ and $\alpha_1 = (12,18,21,12,15)$, then

$$f(u_1,\alpha_1) = \min\{f((1,1,0);16,30,31,22,23) + f((1,0,1);\ 26,32,33,24,33)$$
$$= \min\{37,38\}$$
$$= 37$$

If $u_1 = (0,1,0)$ and $\alpha_1 = (6,18,15,15,12)$, then

$$
\begin{aligned}
f(u_1,\alpha_1) &= \min\{f((1,1,0);14,30,29,23,22) + f((0,1,1);\ 20,32,27,27,30) \\
&= \min\{37,38\} \\
&= 37
\end{aligned}
$$

If $u_1 = (0,0,1)$ and $\alpha_1 = (21,21,18,18,27)$, then

$$
\begin{aligned}
f(u_1,\alpha_1) &= \min\{f((1,0,1);29,33,32,26,37) + f((0,1,1);\ 25,33,28,28,35) \\
&= \min\{41,40\} \\
&= 40
\end{aligned}
$$

Tracing back to Stage 0:

$$
\begin{aligned}
f(u_0,\alpha_0) &= \min\{f((1,0,0);12,18,21,12,15) + f((0,1,0);\ 6,18,15,15,12) \\
&\quad + f((0,0,1);21,21,18,18,27)\} \\
&= \min\{37,37,40\} \\
&= 37
\end{aligned}
$$

Hence the optimal objective value is 37, and there are two optimal sequences: 1-2-3 and 2-1-3.

2.4 β-Robust Scheduling for Single-Machine Systems

A variation to the ADRSP is the β-robust scheduling problem (β-RSP), where each possible scenario is associated with a probability $\rho(P^\lambda)$. The scheduling objective is to determine a sequence that maximizes the likelihood of achieving system performance no worse than a preferred target level T. The processing time random variable p_i is assumed to have a known mean μ_i and variance σ_i^2. The total flow time of a sequence ξ for a given processing time scenario λ is the same as presented earlier, that is,

$$
\varphi(X,P^\lambda) = \sum_{i=1}^{n}\sum_{k=1}^{n}(n-k+1)p_i^\lambda x_{ik}.
$$

The mean and variance of the total flow time distribution then can expressed as

$$
\bar{\varphi}(X) = \sum_{i=1}^{n}\sum_{k=1}^{n}(n-k+1)\mu_i x_{ik}
$$

$$
\sigma^2(\varphi(X)) = \sum_{i=1}^{n}\sum_{k=1}^{n}(n-k+1)^2\sigma_i^2 x_{ik}
$$

The total flow time can be assumed to follow a normal distribution (by invoking the central limit theorem), and the standard normal statistic can be used in the β-RSP formulation (see Daniels and Carrillo, 1997). The quality of the normal approximation of the $\varphi(X, P^\lambda)$ depends on the number of jobs and on the characteristics of the distributions of the individual processing times. A formulation of the β-robust single-machine scheduling problem is as follows:

$$\beta\text{-RSP: } \max \left\{ \frac{T - \sum_{i=1}^{n} \sum_{k=1}^{n} (n - k + 1)\mu_i x_{ik}}{\sum_{i=1}^{n} \sum_{k=1}^{n} (n - k + 1)^2 \sigma_i^2 x_{ik}} \right\} \tag{2.8}$$

subject to

$$\sum_{k=1}^{n} x_{ik} = 1, \quad i = 1, 2, \ldots, n \tag{2.9}$$

$$\sum_{i=1}^{n} x_{ik} = 1, \quad k = 1, 2, \ldots, n \tag{2.10}$$

$$x_{ik} \in \{0, 1\}, \quad i = 1, 2, \ldots, n, \quad k = 1, 2, \ldots, n \tag{2.11}$$

From the problem formulation, it is evident that β-RSP recognizes the effect of both the expectation and variance of the total flow time and seeks to optimize both. The β-RSP also can be shown to be an NP-hard problem (see Daniels and Carrillo, 1997). Daniels and Carillo (1997) also have presented some important properties of β-robust schedules that are useful in developing solution procedures. We highlight these next.

2.4.1 Dominance Properties of β-Robust Schedules

Proposition 2.4. Let $\bar{\varphi}^*$ represent the expected flow time associated with the SPT schedule.

1. If $\mu_i \leq \mu_j$ and $\sigma_i^2 \leq \sigma_j^2$, then there exists a β-robust schedule in which job i precedes j for any $T \geq \bar{\varphi}^*$.
2. If $\mu_i \leq \mu_j$ and $\sigma_i^2 \leq \sigma_j^2$, then there exists a β-robust schedule in which job i precedes j for any $T < \bar{\varphi}^*$.

Note that once satisfied, these properties help in reducing the possible sequences of jobs that need to be explored in order to obtain an optimal solution.

2.4.2 Solution Approaches for β-RSP

There are two solution approaches to solve the β-RSP problem. A branch-and-bound procedure can be developed to solve the problem optimally, whereas an approximate solution can be obtained by using a β-heuristic.

2.4.2.1 Branch-and-Bound Algorithm for β-RSP

A standard branch-and-bound procedure can be developed to solve the β-RSP. As noted earlier, the number of schedules for explicit evaluation can be reduced significantly by using results of Proposition 2.4. The search is further improved by computing appropriate bounds as follows:

Let ξ_p denote a partial sequence generated at the pth level of the tree with p jobs assigned to the first p positions in the sequence, and let ξ_p' denote the complete sequence constructed from ξ_p followed by remaining $n - p$ jobs sequenced in nondecreasing order of μ_i. A lower bound on the expected value of the flow time of any sequence starting with ξ_p then is given by

$$LB_{\bar{\varphi}}(\xi_p) = \sum_{k=1}^{n}(n - k + 1)\mu_{\xi'_{p|k|}}$$

Similarly, when $T \geq \bar{\varphi}^*$, a lower bound on the flow time variance of any schedule ξ_p'' constructed from ξ_p and followed by the $n - p$ jobs arranged in the nondecreasing order of σ_i^2 is given by

$$LB_{\sigma^2[\varphi]}(\xi_p) = \sum_{k=1}^{n}(n - k + 1)^2\sigma_{\xi''_{p|k|}}^2$$

When $T < \bar{\varphi}^*$, the preceding expression represents an upper bound on the flow time variance, and the $n - p$ jobs are arranged in the nonincreasing order of σ_i^2 (follows from Proposition 2.4). An upper bound associated with ξ_p, on the objective function value of β-RSP (2.8) can be determined by using the above two expressions.

$$UB(\xi_p) = [T - LB_{\bar{\varphi}}(\xi_p)]/\sqrt{LB_{\sigma^2[\varphi]}(\xi_p)}$$

Thus, the dominance property and the bounding method can be incorporated into the branch-and-bound algorithm for the β-RSP problem.

2.4.2.2 Heuristic Approach for β-RSP (β-heuristic)

A β-heuristic approach can be used to determine a sequence that provides a lower bound on the probability of achieving a flow time performance level no worse than the target level of T. It is based on generating discrete processing time scenarios that are likely to yield flow time performance near the β-confidence level for any sequence. For a given control parameter value of α, the processing times are determined as

$$p_i = \mu_i + (z_\beta/\alpha)\sqrt{\sigma_i^2}, \quad i = 1, 2, \ldots, n$$

where z_β is defined as

$$z_\beta = [T - \bar{\varphi}^*]/\sqrt{\sigma^2[\varphi(SPT)]}.$$

SPT schedules are constructed for different values of the control parameter α, and they are evaluated using the objective function (2.8). The best value obtained gives the desired lower bound. The quality of the solution obtained depends on the number of values of α used. Daniels and Carrillo (1997) show that good-quality solutions can be obtained by using just a few values of α.

2.4.3 Extensions of the β-RSP

2.4.3.1 Variance Reduction

Processing time uncertainty can be reduced by reformulating the β-RSP to include resource-allocation decisions. A reduction in processing time uncertainty subsequently will amount to a reduced variance of the performance measure (total flow time). Assuming that a single resource of limited supply R exists, the variance present in the processing times of a job decreases as the resource is being consumed by the job. The variance present in the processing time of a job i is given by $\sigma_i^2 = \bar{\sigma}_i^2 - a_i g_i$, where $\bar{\sigma}_i^2$ denotes the normal variance, a_i denotes the rate at which application of the resource reduces the processing time variance of job i, and g_i is the amount of resource allocated to job i. There is also a lower bound $\underline{\sigma}_i^2$ of the processing time variance that is $\underline{\sigma}_i^2 \le \sigma_i^2 \le \bar{\sigma}_i^2$. With T again representing the given target level of performance, a β-RSP with variance reduction (β-RSPVR) can be formulated to determine the optimal sequence and the allocation of resources among the n jobs simultaneously.

$$\beta\text{-RSPVR: max } \frac{T - \sum_{i=1}^{n} \sum_{k=1}^{n} (n - k + 1)\mu_i x_{ik}}{\sqrt{\sum_{i=1}^{n} \sum_{k=1}^{n} (n - k + 1)^2 (\bar{\sigma}_i^2 - a_i g_i) x_{ik}}}$$

subject to

$$\sum_{i=1}^{n} g_i \le R$$

$$0 \le g_i \le \frac{\bar{\sigma}_i^2 - \underline{\sigma}_i^2}{a_i} \quad i = 1, 2, \ldots, n$$

and constraints (2.9), (2.10), and (2.11).

The β-RSPVR maximizes the probability of achieving the target level T while also limiting the amount of resource allocated among the n jobs. β-RSPVR is the generic form of β-RSP and hence is also NP-hard.

The dominance properties, similar to those for the β-RSP, also can be derived for the β-RSPVR (see Daniels and Carrillo, 1997). These are as follows:

Proposition 2.5. Let $\bar{\varphi}^*$ represent the expected flow time associated with the SPT schedule.

1. If $\mu_i \leq \mu_j$, $\bar{\sigma}_i^2 \leq \bar{\sigma}_j^2$, $\underline{\sigma}_i^2 \leq \underline{\sigma}_j^2$, and $a_i \geq a_j$, then there exists an optimal solution to β-RSPVR in which job i precedes j for any $T \geq \bar{\varphi}^*$.
2. If $\mu_i \leq \mu_j$ and $\sigma_i^2 \geq \sigma_j^2$, then there exists an optimal solution to β-RSPVR in which job i precedes j for any $T < \bar{\varphi}^*$.

2.4.4 Solution Approaches for β-RSPVR

2.4.4.1 Branch-and-Bound Algorithm for β-RSPVR

The branch-and-bound algorithm for the β-RSPVR can be developed using Proposition 2.5 and a bounding approach very similar to that for the β-RSP presented earlier in Section 2.4.2.1. For more details, see Daniels and Carrillo (1997).

2.4.4.2 Heuristic approach for β-RSPVR (decomposition heuristic)

The decomposition heuristic used for the β-RSPVR consists primarily of the following two steps: (1) generation of an initial schedule ξ_1 by using the β-heuristic (described earlier in Section 2.4.2.2) with the processing time variances of all the jobs set to their corresponding upper bound values, $\sigma_i^2 = \bar{\sigma}_i^2$, $i = 1, 2, \ldots, n$, and (2) evaluation of the sequence ξ_1 to determine the associated optimal allocation of resources by solving β-RSPVR for the value of x_{ik} fixed in accordance with this sequence. However, the processing time variances obtained for this optimal allocation of resources might be different from those used to construct ξ_1, and hence, the variance values are updated and the steps (1) and (2) are repeated until a previously generated sequence is encountered because further search would result in cycling.

2.5 Robust Scheduling for Two-Machine Flow Shops

We now extend our analysis from single-machine systems to a more complicated machine configuration, namely, a flow shop. We consider the simplest flow-line arrangement consisting of only two machines and permit only permutation sequences. The performance measure of interest, $\varphi(\xi, P^\lambda)$, is the makespan, which can be determined by using Johnson's algorithm (see Johnson, 1954), and the corresponding optimal makespan is denoted by z^λ.

A formulation of the two-machine flow-shop ADRS problem, designated *TM-ADRSP,* is given below (Kouvelis, Daniels, and Vairaktarakis, 2000):

TM-ADRSP: $\min H$

subject to:

$$\sum_{i=1}^{n} p_{i2}^{\lambda} x_{in} + B_n^{\lambda} \leq H + z^{\lambda}, \quad \lambda \in \Lambda$$

$$\sum_{i=1}^{n} \sum_{l=1}^{k} p_{i1}^{\lambda} x_{il} \leq B_k^{\lambda}, \quad k = 1, 2, \ldots, n; \; \lambda \in \Lambda$$

$$B_k^{\lambda} + \sum_{i=1}^{n} p_{i2}^{\lambda} x_{ik} \leq B_{k+1}^{\lambda}, \quad k = 1, 2, \ldots, n-1; \; \lambda \in \Lambda$$

$$\sum_{k=1}^{n} x_{ik} = 1, \quad i = 1, 2, \ldots, n$$

$$\sum_{i=1}^{n} x_{ik} = 1, \quad k = 1, 2, \ldots, n$$

$$x_{ik} \in \{0, 1\} \quad i = 1, 2, \ldots, n \quad k = 1, 2, \ldots, n$$

where p_{ij}^{λ} = processing time of job i on machine j in scenario λ

$$x_{ik} = \begin{cases} 1, & \text{if } \xi_{[k]} = i \\ 0, & \text{otherwise} \end{cases}$$

B_k^{λ} = start time of job $\xi_{[k]}$ on machine 2 given processing time scenario λ.

The TM-ADRSP is NP-hard (see Kouvelis, Daniels, and Vairaktarakis, 2000). The problem formulation above is valid for both the discrete time scenario and the continuous processing time interval cases. Next, we highlight some of the important properties associated with each case (see Kouvelis, Daniels, and Vairaktarakis, 2000).

2.5.1 Properties of Two-Machine Flow-Shop Robust Schedules

2.5.1.1 Discrete Processing Time Scenarios

Proposition 2.6. If for two jobs i and j, $p_{i1}^{\lambda} \leq p_{j1}^{\lambda}$ and $p_{i2}^{\lambda} \geq p_{j2}^{\lambda}$ for all $\lambda \in \Lambda$, then there exists an absolute deviation robust schedule in which job i precedes job j.

Proposition 2.7. If for two jobs i and j, $\min\{p_{i1}^{\lambda}, p_{j2}^{\lambda}\} \leq \min\{p_{j1}^{\lambda}, p_{i2}^{\lambda}\}$ for all $\lambda \in \Lambda$, then the absolute deviation robust schedule can be determined without explicitly considering schedules in which job j immediately precedes job i.

2.5.1.2 Continuous Processing Time Intervals

Given a sequence ξ and a processing time scenario λ, job $\xi_{[k]}$ is called a *critical job* if it is processed on machine 2 immediately after its completion on machine 1. Let C_λ^ξ be the index set of critical jobs: $C_\lambda^\xi = \{k: \xi_{[k]} \text{ is a } critical \ job \text{ for sequence } \xi \text{ and scenario } \lambda\}$.

Proposition 2.8.

1. For any sequence ξ and the makespan performance criterion for the two-machine flow shop, the worst-case absolute deviation scenario λ_0^ξ belongs to the set of extreme point scenarios.
2. For the worst-case absolute deviation scenario λ_0^ξ with respect to sequence ξ:

$$p_{ij}^{\lambda_0^\xi} = \begin{cases} \overline{p}_{ij}, & \text{if } (i \in V_1^\xi \cup \{\xi_{[i_0]}\} \text{ and } j = 1) \text{ or } (i \in V_2^\xi \cup \{\xi_{[i_0]}\} \text{ and } j = 2) \\ \underline{p}_{ij}, & \text{if } (i \in V_2^\xi \text{ and } j = 1) \text{ or } (i \in V_1^\xi \text{ and } j = 2) \end{cases}$$

where $\xi_{[i_0]}$ is the last critical job in sequence ξ for processing time scenario λ_0^ξ, and V_1^ξ and V_2^ξ are defined as $V_1^\xi = \{\xi_{[k]} : k < i_0\}$ and $V_2^\xi = \{\xi_{[k]} : k > i_0\}$, respectively.

2.5.2 Solution Approaches for the TM-ADRSP

2.5.2.1 Branch-and-Bound Algorithm for TM-ADRSP

Exact solutions for the TM-ADRSP can be found by using a branch-and-bound procedure for the discrete processing time scenarios and continuous processing time intervals cases (see Kouvelis, Daniels, and Vairaktarakis, 2000). A brief discussion of the procedure for both cases is provided below.

2.5.2.1.1 DISCRETE PROCESSING TIME SCENARIOS. The worst-case scenario for any given sequence can be determined as follows: For a $\lambda \in \Lambda$, the makespan of a given sequence is computed and the optimal makespan for the corresponding λ is determined using Johnson's algorithm. The difference between the makespan values is noted, and the iterations are repeated for all the other scenarios. At the end of $|\Lambda|$ iterations (total number of processing time scenarios), the largest value obtained represents the worst-case deviation from optimality for the given sequence.

The process of determining a lower bound on the worst-case deviation from optimality of any partial schedule is similar to the preceding procedure and also requires $|\Lambda|$ iterations. Thus, for any partial schedule $1, 2, \ldots, p$ at iteration λ of the bounding process, a new schedule is constructed by arranging the remaining

$n - p$ jobs in Johnson's order and appending that to the partial schedule. The difference in the makespan value of the optimal schedule for scenario λ and the makespan value of the constructed schedule represents a lower bound on the absolute deviation from optimality that can be realized given the partial sequence and scenario λ. The maximum lower bound so obtained by repeating the process for all $\lambda \in \Lambda$ represents the minimum worst-case deviation from optimality that can be realized given the partial sequence and scenario λ. This bounding approach along with dominant properties presented in Propositions 2.6 and 2.7 can be incorporated in the branch-and-bound algorithm, similar to the one described in Section 2.3.2.1 to optimally solve the TM-ADRSP.

2.5.2.1.2 CONTINUOUS PROCESSING TIME INTERVALS. For this case, the worst-case deviation from optimality, for a given sequence, is determined as follows: There are n iterations in this procedure. Iteration k of the procedure corresponds to the job occupying the kth position in the sequence. At iteration k, the kth job is assumed to be the last job, and the processing times for all the other jobs are set using Proposition 2.8. The makespan of the given sequence is computed and compared with that of the optimal makespan, determined using Johnson's algorithm for the preceding set of processing times. The deviation from optimality is recorded, and at the end of n iterations, the worst-case absolute deviation from optimality for the given sequence is obtained.

A lower bound for a given partial sequence can be developed using the preceding procedure. Consider any partial schedule $1, 2, \ldots, p$ and the positions of the remaining unscheduled $n - p$ jobs as yet unknown. At iteration k (job occupying the kth position in the sequence) of the bounding process, the kth job is assumed as the critical job. The processing times of all the jobs are determined using Proposition 2.8. Given this processing time scenario and job k as the last job in the final schedule, a lower bound on the makespan associated with the partial schedule is computed as follows:

$$LB = \sum_{i=1}^{k} \bar{p}_{i1} + \sum_{i=k}^{n} \bar{p}_{i2}$$

The corresponding optimal makespan is determined using Johnson's algorithm. The deviation between these two values is retained if the difference is greater than the largest difference from the earlier $k - 1$ iterations. After n iterations, a lower bound on the worst-case deviation from optimality for the given partial sequence is obtained.

2.5.2.2 Heuristic Approaches for TM-ADRSP
Since the TM-ADRSP is quite complex to solve to optimality, heuristic solution approaches are available to obtain approximate and quick solutions. Separate

heuristic approaches are available for the discrete processing time scenario and continuous processing time interval cases (see Kouvelis, Daniels, and Vairaktarakis, 2000).

2.5.2.2.1 DISCRETE PROCESSING TIME SCENARIOS. This procedure is based on the notion that the robust schedule will be structurally similar to at least one of the schedules that is optimal with respect to a particular scenario. This procedure requires $|\Lambda|$ iterations. Starting with a scenario λ, the Johnson's sequence is constructed and the corresponding worst-case performance is determined using the procedure described in Section 2.5.2.1.1. This sequence then is improved by considering $n(n-1)$ insertions and $n(n-1)/2$ pairwise interchanges to generate newer sequences. The worst-case deviation from optimality then is determined for each sequence using procedure described in Section 2.5.2.1.1. If no improvement is realized, the procedure advances to the next scenario $\lambda + 1$; otherwise, the alternative sequence yielding the lowest objective value becomes the incumbent. The entire process then is repeated for $|\Lambda|$ iterations, and at the end of it, an approximate solution that minimizes the worst-case deviation from optimality is obtained.

2.5.2.2.2 CONTINUOUS PROCESSING TIME INTERVALS. The approximate solution approach for this case is based on generating an initial sequence by using the last critical job and improving the solution via the procedure described in Section 2.5.2.1.2. Let L and R be defined as the sets of jobs that occupy sequence positions before and after the last critical job, respectively. A sequence then can be constructed by arranging the jobs $i \in L$ in Johnson's order with respect to $\bar{p}_{i1}$ and $\underline{p}_{i2}$, followed by jobs $i \in R$ arranged in Johnson's order with respect to $\underline{p}_{i1}$ and $\bar{p}_{i2}$. The worst-case performance of this initial sequence is determined using the procedure described in Section 2.5.2.1.2. As before, the sequence is improved by considering $n(n-1)$ insertions and $n(n-1)/2$ pairwise interchanges to generate newer sequences. The worst-case deviation from optimality then is determined for each sequence using the procedure described in Section 2.5.2.1.2. If no improvement is realized, the procedure terminates. Otherwise, the alternative sequence yielding the lowest objective value becomes the incumbent. The entire process then is repeated until there is no improvement in order to determine an approximate solution that minimizes the worst-case deviation from optimality.

2.6 Concluding Remarks

In this chapter we have reviewed in detail the modeling and solution approaches that are available in the literature to identify robust schedules in a variable scheduling environments where variability is caused by randomness in the job processing times. The scheduling environments that are considered include single-machine and two-machine flow-shop systems. Both exact and heuristic methodologies have been presented. The exact methodology uses the branch-and-bound approach that relies on some inherent properties of the problem at hand.

3 Expectation-Variance Analysis in Stochastic Multiobjective Scheduling

3.1 Introduction

Stochastic multiobjective scheduling always has been an interesting and challenging field because the task of evaluating a schedule under multiple criteria is highly complex. The stochastic nature of the scheduling problems in our discussion is due to the randomness of the job processing times. A generalized form of a stochastic multiobjective scheduling problem when the processing times are probabilistic can be stated as follows:

$$\min \left\{ f(\xi, C(\omega)) \middle| \xi \in \Pi \omega \in \Omega \right\}$$

where $f(\cdot)$ denotes a scheduling criterion vector function, and Π denotes a set of feasible schedules ξ. The element $C(\omega)$ is a random matrix that consists of $c_{ij}(\omega)$, the random processing time of a job j on machine i, and is defined on some probability space Ω.

The multiple criteria on which we are focusing are expected value and the variance of a performance measure of evaluation. The scheduler's objective in this case is to determine a schedule that is best in terms of the mean and variance of the performance measure. The significance of considering expectation and variance for the objective function of stochastic scheduling problems is that the focus on the expectation alone might not reflect the scheduler's risk attitude, and to account for risk accurately, a scheduler rather must strive to maximize the expected utility function. Since the utility function might be difficult to define, the other viable alternative is to determine stochastically optimal and efficient sequences. However, determining stochastically optimal or efficient sequences is technically complicated and not of much relevance in practice. Hence, analyzing the schedules based on their efficiencies with respect to the expected value and variance of the performance measure seems a feasible and viable option (De et al., 1992).

A stochastic multiobjective scheduling problem with the expectation and variance functions can take the following form:

$$\min \left\{ \left(\begin{array}{c} E[f(\xi, C(\omega))] \\ \text{var}[f(\xi, C(\omega))] \end{array} \right) \middle| \xi \in \Pi, \omega \in \Omega \right\}$$

For stochastic multiobjective scheduling problems, it is unlikely that a solution will be identified that is optimal in all the criteria simultaneously because a solution that is good for one objective may not be good for the other objective or objectives. Furthermore, the solution also depends on the weights or importance factors that a production manager or scheduler assigns to each of the objectives. Different managers will have different priorities, and hence there is no one optimal solution that would be universally acceptable. Predominantly, in multicriteria scheduling problems, the solution primarily constitutes a set from which the user can select his or her schedule that appropriately suits the desired preferences.

3.2 Expectation-Variance-Efficient Sequences/Nondominated Schedules

An attempt in this regard has been to use the notion of identifying a set of schedules called *nondominated schedules*. Since there is more than one objective, this set contains the solutions that are reasonably good in terms of one criterion or the other. Incidentally, this set also includes all the Pareto-optimal solutions with respect to multiple objectives. Other solutions that are worse in terms of both the criteria with respect to the schedules in the nondominated set are discarded. Various efficient heuristics for generating this nondominated set for multiobjective scheduling problems have been developed and are available in the literature (see Jung et al., 1990; Morizawa et al., 1993; Rajendran, 1995; Murata et al., 1996). However, this set of nondominated schedules might contain hundreds of schedules, and it becomes hard for the decision maker to select a preferred schedule among so many schedules by comparing their criterion vector values with each other. Hence the major task in stochastic multiobjective scheduling is to determine a methodology for finding a preferred schedule from among the set of nondominated schedules based on certain user inputs.

On the other hand, one also can find use of the term *expectation-variance-efficient sequences* when identifying schedules that are efficient in terms of the mean and variance of the performance measure (De et al., 1992). This set of expectation-variance-efficient sequences is analogous to the set of nondominated schedules.

In the following sections we first discuss the methodologies that are available to identify the set of expectation-variance-efficient sequences. Then we present different methodologies that are available for identifying the preferred schedule from a nondominated set of schedules for various scheduling environments. The multiple objectives in all our discussion are the expectation and variance of the performance measure. The expectation and variance objectives also can be referred to as "conflicting" objectives, because most of the time a good solution with respect to the expected value of the performance measure is quite poor in terms of its variance. We begin this discussion on single-machine scheduling problems with total flow time as the criterion.

3.3 Identification of Expectation-Variance-Efficient Sequences

The problem considered is of static scheduling of n jobs with random processing times on a single machine with no preemption. The objective of interest is to minimize the total flow time of all jobs.

Consider the following notation:

j = job index

N = set of all jobs

p_j = processing time of job j, which is a random variable with expectation μ_j and variance σ_j^2

$F_j(\xi)$ = flow time of job j in schedule ξ

The total flow time then is given by $F(\xi) = \sum_{j \in N} F_j(\xi)$.

It also could be expressed as $F(\xi) = \sum_{k \in N} k p_{[k]}$, where $[k]$ denotes the index of the job in the kth last position in schedule ξ.

A sequence ξ is termed as *expectation-variance-efficient* (EV-efficient) if there exists no other sequence ξ' such that $EF(\xi') \leq EF(\xi)$ and $VF(\xi') \leq VF(\xi)$, with at least one of the inequalities holding strictly. EF and VF are the expected value and variance of the total flow time, respectively.

The expected value and variance of the total flow time can be expressed as $EF(\xi) = \sum_{k \in N} k \mu_{[k]}$ and $VF(\xi) = \sum_{k \in N} k^2 \sigma_{[k]}^2$, respectively.

3.3.1 Approaches for Identifying EV-Efficient Sequences

Two different approaches are available to determine EV-efficient sequences for the preceding flow time problem. One approach is an extension of the standard dynamic programming–based method owing to Held and Karp (1962) (designated as *EV-DP*). The other approach solves the problem as a linear assignment problem subject to a single side constraint (designated as *EV-LAP, bicriteria assignment problem*).

3.3.1.1 Dynamic Programming Approach (EV-DP)

In this approach, sequences are generated by scheduling the N jobs successively from the last position onward. ξ_{k-1} denotes a partial sequence thus generated by assigning $(k-1)$ jobs to the last $(k-1)$ positions, and S_{k-1} denotes the set of $(k-1)$ jobs. It can be shown that the completion of ξ_{k-1} cannot lead to an EV-efficient sequence unless ξ_{k-1} is itself EV-efficient with respect to the jobs in S_{k-1}. Thus, as we proceed to construct sequences from the last position, it is necessary to consider and retain only EV-efficient sequences. Hence a partial sequence ξ_{k-1} can be expanded by scheduling a job in the kth last position to form sequence ξ_k. Hence the expected value and variance of this sequence ξ_k is given by $EF(\xi_k) = EF(\xi_{k-1}) + k\mu_j$ and $VF(\xi_k) = VF(\xi_{k-1}) + k^2\sigma_j^2$. Thus a dynamic programming approach can be developed to solve the problem incorporating the preceding principle. A detailed explanation of the DP approach with an illustrative example can be found in De et al. (1992).

Notes about the EV-DP approach:

1. The computational effort expended in the worst case could be quite large.
2. The number of sequences to be considered over the n stages is bounded by

$$\sum_{k=1}^{n}(n-k+1)\rho(n,k-1)\theta_{k-1}$$

   where $\rho(n,k-1)$ represents the number of ways in which $(k-1)$ jobs can be selected out of n, and θ_{k-1} represents the maximum number of distinct EV-efficient partial sequences that can be formed with any given set of $(k-1)$ jobs.
3. The computational effort can be reduced by exploiting dominance relations between the jobs. A job j dominates j' if $\mu_j \leq \mu_{j'}$ and $\sigma_j^2 \leq \sigma_{j'}^2$, with one of the equalities being strict. If so, job j must precede j' in any EV-efficient sequence.

3.3.1.2 Linear-Assignment-Problem Approach (EV-LAP)

Let $e(j,k)$ and $v(j,k)$ denote the contributions of job j to the $EF(\xi)$ and $VF(\xi)$ functions if it is assigned to the kth last position in ξ, and they are, respectively, given by $e(k,j) = k\mu_j$ and $v(k,j) = k^2\sigma_j^2$. For a feasible solution, set A represents a complete list of assignment pairs (j,k), where $j,k \in N$. The problem of finding EV-efficient sequences is then equivalent to a bicriteria assignment problem where we try to minimize both $\sum_{(j,k)\in A} e(k,j)$ and $\sum_{(j,k)\in A} v(k,j)$ simultaneously. Let ξ^0 denote the sequence with the smallest total variance (VF) among all the sequences that achieve the optimal expected value (EF). Let ξ^∞ denote the sequence with the smallest total expected value (EF) among all the sequences that achieve the optimal total variance (VF).

The solution procedure is as follows:

Step 1. Set $VF^0 = VF(\xi^0)$, and iterate over t, starting at $t = 1$.
Step 2. Find sequence ξ that minimizes $\sum_{(j,k)\in A} e(k,j)$ subject to $\sum_{(j,k)\in A} v(k,j) \leq VF^{t-1} - \delta$, where δ is an appropriately small constant.
Step 3. If $EF(\xi) = EF(\xi^\infty)$, then stop; else go to step 4.

Step 4. Find sequence $\underline{\xi}^t$ that minimizes $\sum_{(j,k)\in A} v(k,j)$ subject to $\sum_{(j,k)\in A}$ $e(k,j) \leq EF(\xi)$, and assign $\underline{\xi}^t$ to be EV-efficient. Set $\underline{VF^t} = VF(\underline{\xi}^t)$, set $t = t + 1$, and go to step 2.

Notes about EV-LAP approach:

1. The algorithm terminates after $(\theta_n - 1)$ iterations, and at each iteration, except for the last, two assignment problems with a single constraint (APSC) are solved, where θ_n represents the cardinality of a completely representative set of EV-efficient sequences. Thus most of the computational effort is expended on solving $(2\theta - 1)$ APSCs.
2. The worst-case complexities are exponential for APSC algorithms.

3.4 Identification of Extreme EV-Efficient Sequences

An EV-efficient sequence ξ is termed *extreme-expectation-variance-efficient* or *XEV-efficient* if there exists an $\alpha \in [0,1]$ for which ξ minimizes $Z_a = \alpha EF + (1-\alpha)VF$. The XEV-efficient sequences are determined by solving the problem as a bicriteria transportation problem (XEV-LAP).

3.4.1 Linear-Assignment-Problem Approach (XEV-LAP)

The algorithmic procedure for the XEV-LP approach is as follows:

Step 1. Set $L = \{(\xi^0, \xi^\infty)\}$, and iterate over t (starting at $t = 1$).
Step 2. If L is empty, stop; else, remove a pair (ξ^c, ξ^d) from L and compute $\alpha = [VF(\xi^c) - VF(\xi^d)]/[EF(\xi^d) - EF(\xi^c) + VF(\xi^c) - VF(\xi^d)]$.
Step 3. Find a sequence $\underline{\xi}^t$ that, among all sequences that minimize $Z_\alpha(\cdot)$, has the minimum $EF(\cdot)$.
Step 4. Check if $(EF(\underline{\xi}^t), VF(\underline{\xi}^t))$ is distinct from $(EF(\xi^c), VF(\xi^c))$ and $(EF(\xi^d), VF(\xi^d))$. If so, record $\underline{\xi}^t$ to be XEV-efficient, add $(\xi^c, \underline{\xi}^t)$ and $(\underline{\xi}^t, \xi^d)$ to L, begin a new iteration $(t = t + 1)$, and go to step 2; else, begin a new iteration $(t = t+1)$, and go to step 2.

Note about XEV-LP approach:
If ϕ_n is the cardinality of a completely representative set of XEV-efficient sequences, the XEV-LAP terminates after $(2\phi_n - 3)$ iterations, and at each iteration, it requires solution of an assignment problem. An assignment problem can be solved in $O(n^3)$ time, and since $\phi_n \leq \theta_n$, it is concluded that determining the set of XEV-efficient sequences would be easier than determining the set of EV-efficient sequences.

3.5 Preferred Schedule for Bicriteria Single-Machine Scheduling

As presented earlier, there are several ways of identifying a set of efficient sequences or nondominated schedules. In this section we delve into the different methodologies that can be used to select a preferred schedule from this set of schedules. The idea is to assist the scheduler in selecting a preferred schedule from among a set of nondominated schedules N through an interactive system termed the *interactive stochastic multiobjective scheduling system* (ISMSS) by Nagasawa and Shing (1998).

The objective is similar to that of the earlier cases, that is, to schedule jobs with random processing times on a single machine such that both the expected value and the variance of the total flow time are minimized.

The primary stochastic multiobjective scheduling problem could be expressed as

$$P1: \min \left\{ \left. \begin{pmatrix} E[F(\xi)] \\ \text{var}[F(\xi)] \end{pmatrix} \right| \xi \in \Pi \right\}$$

where $F(\xi)$, $E[F(\xi)]$, and $\text{var}[F(\xi)]$, respectively, denote the total flow time, the expected value, and the variance of the total flow time associated with schedule ξ belonging to the set of permutation schedules Π.

Jung et al. (1990) proposed a heuristic procedure based on a pairwise-job-interchange method to determine an approximate set of nondominated schedules to the problem P1. As mentioned earlier, the cardinality of the set N is normally quite large, and it becomes imperative to determine an efficient way to select a preferred schedule from the set based on the user's requirements. Prior to that, some new categories associated with identifying the preferred schedule from the set of nondominated schedules are presented below.

3.5.1 Upperward 100α Percentile Minimum Schedule

The solution obtained by minimizing the expected value $E[F(\xi)]$ alone allows the total flow time to exceed the expected value with a probability of 50%. However, it would most likely be desirable for the probability of the total flow time exceeding the expected value not to be very large. If y is the upperward 100α percentile, that is, the total flow time exceeds y with a probability of α, the problem could be formulated as

$$P2: \min \left\{ y \middle| \Pr[F(\xi) \leq y] \geq 1 - \alpha, \xi \in \Pi \right\}$$

If the total flow time is assumed to follow a normal distribution with mean $E[F(\xi)]$ and variance $\text{var}[F(\xi)]$, the problem could be reformulated as

$$P3: \min \left\{ E[F(\xi)] + u_\alpha \sqrt{\text{var}[F(\xi)]} \middle| \xi \in \Pi \right\}$$

where $\Phi(u_\alpha) = 1 - \alpha$, $\Phi(\cdot)$ is the cumulative distribution function of a standard normal distribution, and u_α is the $100\alpha\%$ point in the standard normal distribution. The optimal schedule to this problem is called as the *upperward*

100α percentile minimum schedule and is denoted by $\xi(\alpha)$. It can be obtained by searching the nondominated schedules in N.

3.5.2 Combined Nondominated Schedules

In order to also guarantee that the $\mathrm{var}[F(\xi)]$ is minimized for a given α, the scheduling problem could be reformulated as follows:

$$
\text{P4: } \min \left\{ \begin{array}{c} \left(\dfrac{E[F(\xi)] + u_\alpha \sqrt{\mathrm{var}[F(\xi)]}}{\sqrt{\mathrm{var}[F(\xi)]}} \right) \\ \xi \in \Pi \end{array} \right\}
$$

The set of nondominated schedules to P4 is called the set of *combined non-dominated schedules* and is denoted by $N_c(\alpha)$. These nondominated schedules can be obtained by computing

$$
u_\alpha(\xi_i) = \min \left\{ \left| \frac{E[F(\xi_j)] - E[F(\xi_i)]}{\sqrt{\mathrm{var}[F(\xi_i)]} - \sqrt{\mathrm{var}[F(\xi_j)]}} \right| \ \mathrm{var}[F(\xi_i)] > \mathrm{var}[F(\xi_j)], \xi_j \in N \right\}
$$

where the ξ_j's and ξ_i's are schedules in N, and

$$
N_c(\alpha) = N / \{ \xi_i | u_\alpha(\xi_i) \leq u_\alpha, \xi_i \in N \}
$$

3.5.3 Combined Upperward 100α Percentile Minimum Schedules

Another formulation of the same scheduling problem for which we wish to minimize both y and α could be

$$
\text{P5: } \min \left\{ \left(\begin{array}{c} y \\ \alpha \end{array} \right) \ \middle| \ \Pr[F(\xi) \leq y] \geq 1 - \alpha, \xi \in \Pi \right\}
$$

If the distribution of the total flow time similarly is assumed to be normal, P5 can be rewritten as

$$
\text{P6: } \min \left\{ \left(\begin{array}{c} E[F(\xi)] + u_\alpha \sqrt{\mathrm{var}[F(\xi)]} \\ u_\alpha \end{array} \right) \ \middle| \ \xi \in \Pi \right\}
$$

The set of nondominated schedules to P6 is called the combined *upperward 100α percentile minimum schedules* and is denoted by N^α, which can be obtained from N as follows:

$$
N^\alpha = \{ \xi(\alpha) | 0 < \alpha \leq 0.5 \}
$$

3.5.4 Algorithmic Procedure for Preferred Schedule Selection

A stepwise interactive algorithm to determine the preferred schedule can be found in Nagasawa and Shing (1998) and Shing and Nagasawa (1997). The brief steps are as follows:

Step 1. The set of nondominated schedules N is first determined using the heuristic algorithm proposed by Jung et al. (1990). Once N is known, the sets $N_c(\alpha)$ and N^α can be generated.

Step 2. Later, these sets are plotted, and the upper and lower limits of α, α_1 and α_2, respectively, are adjusted to express the set $\{\xi(\alpha) | \alpha_1 \leq \alpha \leq \alpha_2\}$, with a reasonable number of candidate schedules.

Step 3. A preferred schedule can be selected from this set by comparing their criterion vector values.

Note:

The categories discussed earlier to select a preferred schedule for the single-machine bicriteria scheduling problem also can be applied directly to portfolio models that deal with the expected value and the variance of the return on investments. Higher returns generally are associated with higher risks (denoted by variance), and the objective is to select a portfolio that optimizes (or maximizes) the expected value and minimizes the variance of the return based on different priorities of the user. Detailed study on these models can be found in Shing and Nagasawa (1999).

3.6 Preferred Schedule for Bicriteria Parallel-Machine Scheduling

In this environment, there are n jobs that are to be processed in a parallel M-machine system. Two different cases can be considered: (1) the job assignment is given in advance, and (2) the job assignment is not fixed. For case 1, if we evaluate every schedule for the individual machine, the procedure detailed in the preceding section can be applied directly (*individual evaluation*). However, this individual evaluation may not be a good measure for the overall system, and hence it is also necessary to study the system as a whole (*overall evaluation*). For case 2, only the overall evaluation is considered because the individual evaluation is highly complex.

3.6.1 Fixed-Job-Assignment Case

3.6.1.1 Individual Evaluation

The individual evaluation of machine i in a bicriteria parallel-machine scheduling problem can be formulated as follows:

$$P7^{(i)}: \min \left\{ \begin{pmatrix} E[F(\xi^{(i)})] \\ \text{var}[F(\xi^{(i)})] \end{pmatrix} \middle| \xi^{(i)} \in \Pi^{(i)}(B^{(i)}), B^{(i)} \text{ is given} \right\}, \quad i = 1, \ldots, M$$

where $B^{(i)}$ denotes the set of jobs already assigned to machine i, and $\Pi^{(i)}(B^{(i)})$ denotes the set of feasible schedules $\xi^{(i)}$ generated from $B^{(i)}$. Since $B^{(i)}$ is known for all machines, $i = 1, \ldots, M$, the set of nondominated schedules for each machine $N^{(i)}$ can be found by applying the heuristic method of Jung et al. (1990) to each machine.

3.6.1.2 Overall Evaluation

The formulation for the overall evaluation of a parallel M-machine scheduling problem is as follows:

$$
\text{P8: min} \left\{ \left. \left(\begin{array}{c} \sum_{i=1}^{M} E[F(\xi^{(i)})] \\ \sum_{i=1}^{M} \text{var}[F(\xi^{(i)})] \end{array} \right) \right| \xi^{(i)} \in \Pi^{(i)}(B^{(i)}), B^{(i)} \text{ is given}, \quad i = 1, \ldots, M \right\}
$$

$\Pi^{(i)}(B^{(i)})$ in P8 can be replaced by $N^{(i)}(B^{(i)})$ because any dominated schedule to P7$^{(i)}$ cannot become any component of nondominated schedules to P8. Furthermore, the objective function and constraints in P8 are all additive; hence it can be decomposed as follows:

$$
\text{P9: min} \left\{ \text{min} \left\{ \left. \left(\begin{array}{c} \sum_{i=1}^{M-1} E[F(\xi^{(i)})] \\ \sum_{i=1}^{M-1} \text{var}[F(\xi^{(i)})] \end{array} \right) \right| \xi^{(i)} \in N^{(i)}, \quad i = 1, \ldots, M-1 \right\} \right.
$$
$$
\left. + \left(\begin{array}{c} E[F(\xi^{(M)})] \\ \text{var}[F(\xi^{(M)})] \end{array} \right) \right| \xi^{(M)} \in N^{(M)} \right\}
$$

The decomposition form allows the original problem to be reduced to smaller-sized problems. An algorithm to determine the set of nondominated schedules to problem P9 is as follows:

Step 1. Determine the set of nondominated schedules for each machine by solving the individual machine problem (by solving M P7$^{(i)}$ problems).

Step 2. A set of feasible candidate schedules can be generated for a two-parallel-machine problem ξ from the sets $N^{(1)}$ and $N^{(2)}$ (the sets of nondominated schedules for machines 1 and 2, respectively) with respect to the two objectives of $\sum_{i=1}^{2} E[F(\xi^{(i)})]$ and $\sum_{i=1}^{2} \text{var}[F(\xi^{(i)})]$.

Step 3. Thus a k-parallel-machine problem can be solved using the sets of nondominated schedules for the $(k-1)$-parallel-machine problem and the machine-k problem in a similar way to step 2.

Step 4. The procedure is continued progressively until the overall M-parallel-machine problem is solved.

3.6.2 General Parallel-Machine Case

Here, the assignments of the jobs to the machines are not fixed. Accordingly, we have the formulation

$$
\text{P10:} \min \left\{ \left. \left(\begin{array}{c} \sum_{i=1}^{M} E[F(\xi^{(i)})] \\ \sum_{i=1}^{M} \text{var}[\xi(\pi^{(i)})] \end{array} \right) \right| \xi^{(i)} \in \Pi^{(i)}(\beta^{(i)}),\ i=1,\ldots,M, \bigcup_{i=1}^{M} \beta^{(i)} = B \right\}
$$

where B denotes a set of jobs to be assigned among M machines, and $\beta^{(i)}$, $i=1,\ldots,M$, represents any decomposition of B such that $\bigcup_{i=1}^{M} \beta^{(i)} = B$ and $\beta^{(i)} \cap \beta^{(j)} = \phi, \forall i \neq j$.

Similar to the fixed-job-assignment case, $\Pi^{(i)}(\beta^{(i)})$ in P10 can be replaced by $N^{(i)}(\beta^{(i)})$, and if the decomposition of $\beta^{(i)}, i=1,\ldots,M$, is given, the set of nondominated schedules for P10 can be derived using the four-step method presented in Section 3.6.1. A final set of nondominated schedules can be determined by evaluating the whole sets of nondominated schedules generated for all possible decompositions of B. This task could be highly cumbersome, and fortunately, as shown in Proposition 3.1 below (see Nagasawa and Shing, 1998), it is sufficient to consider only balanced job assignments, where the difference in the number of jobs assigned to each machine is at most one. Decompositions that cannot provide any nondominated schedule are omitted.

Proposition 3.1. For an N-job, M-parallel machine stochastic bicriteria scheduling problem defined by P10, the set of nondominated schedules can be generated by considering all decompositions $(\beta^{(1)}, \beta^{(2)}, \ldots, \beta^{(M)})$ such that $\bigcup_{i=1}^{M} \beta^{(i)} = B$ and

$$
N_i = \begin{cases} \lfloor N/M \rfloor + 1, & \text{if } i = 1,\ldots,N - \lfloor N/M \rfloor M \\ \lfloor N/M \rfloor, & \text{if } i = N - \lfloor N/M \rfloor M + 1,\ldots,M \end{cases}
$$

where $N_i \equiv |\beta^{(i)}|, i = 1,\ldots,M$

3.6.3 Algorithmic Procedure for Preferred Schedule Selection

The basic idea for selecting the preferred schedule in the parallel-machine case is the same as in the single-machine case. Especially under the overall evaluation case for the fixed-job assignment and general formulations, the single evaluation method can be used without any change. However, the individual evaluation case for the fixed-job assignment, parallel-machine problem requires some minor modifications with the underlying procedure being the same (see Nagasawa and Shing, 1997, 1998).

3.7 Concluding Remarks

In this chapter we have presented the concept of nondominated schedules and EV-efficient sequences in the field of stochastic scheduling. Different approaches to identify this set of schedules were described, and the algorithms to select a preferred schedule from among this schedule set also were discussed. The scheduling environments considered were those of a single machine and parallel machines.

4 Single-Machine Models

4.1 Introduction

In this chapter we consider the problem of sequencing a given number of jobs on a single machine and devise methodologies and develop closed-form expressions (wherever possible) to compute the expectation and variance of various performance measures. Numerical illustrations indicating the significance and applicability of our work are also presented through example problems.

Besides being of significance in their own right, single-machine problems also constitute a good starting point for analyzing more complex scheduling environments of flow shop and job shop. The different performance measures that we consider for the single-machine case can be classified into two categories:

1. *Completion-time-based*, which includes total completion time (total flow time), total weighted completion time, and total weighted discounted completion time
2. *Due-date-based*, which includes total tardiness, total weighted tardiness, total number of tardy jobs, total weighted number of tardy jobs, mean lateness, and maximum lateness.

The processing time of a job is assumed to be a random variable that follows an arbitrary probability distribution.

First, we present the notation that is used in this chapter. Some of this notation has been introduced already in earlier chapters. However, we include it here for the sake of completeness.

n = number of jobs to be scheduled on the machine

All the following are defined for the job located at the jth position of a given sequence.

$P_{[j]}$ = processing time (a random variable)
$\mu_{[j]}$ = mean or expected value of the processing time

$\sigma^2_{[j]} = $ variance of the processing time
$C_{[j]} = $ completion time (a random variable)
$w_{[j]} = $ weight or importance factor
$d_{[j]} = $ due date
$L_{[j]} = $ lateness
$T_{[j]} = $ tardiness
$U_{[j]} = $ unit penalty

$C_{[j]}$ is defined as the time at which a job exits the system after completing its processing on the machine. It is equal to the sum of the processing times of all the jobs that are processed prior to this job on that machine. Hence it is primarily the sum of the waiting time of the job and its processing time:

$$C_{[j]} = \sum_{i=1}^{j-1} P_{[i]} + P_{[j]} = \sum_{i=1}^{j} P_{[i]}$$

The lateness of a job in the jth position is defined as $L_{[j]} = C_{[j]} - d_{[j]}$, which is positive when job $[j]$ is completed late (after its due date) and negative when it is completed early (before its due date).

The tardiness of the job in the jth position is defined as

$$T_{[j]} = \max(0, L_{[j]}) = \max(0, C_{[j]} - d_{[j]})$$

The unit penalty of job in the jth position is defined as

$$U_{[j]} = \begin{cases} 1, & \text{if } C_{[j]} > d_{[j]} \\ 0, & \text{if } C_{[j]} \leq d_{[j]} \end{cases}$$

4.2 Completion-Time-Based Objectives
4.2.1 Total Completion Time

Since we assume that all jobs are available at time zero, the completion time and flow time are identical in our case. Completion time measures the response of the system to individual demands of service and represents the interval of time between the arrival and departure of a job (this interval is also known as the *turnaround time*). The significance of this objective lies in the fact that a minimal total completion time of the jobs will help to maintain a low average in-process inventory (Baker, 1974).

For a given schedule, the total completion time is the sum of the completion times of all the jobs $\sum_{j=1}^{n} C_{[j]} = \sum_{j=1}^{n} \sum_{i=1}^{j} P_{[j]}$. By rearranging and summing up similar terms in this expression, we get

$$\sum_{j=1}^{n} C_{[j]} = \sum_{j=1}^{n} (n + 1 - j) P_{[j]}$$

The expectation and variance of the total completion time are given below.

Table 4.1. *Data Set for Example 4.1*

Job Index	1	2	3	4
Mean	40	20	60	25
Variance	15	15	20	5
Weight	5	4	10	5
Due date	120	80	90	60

Expectation of the total completion time:

$$E\left[\sum_{j=1}^{n} C_{[j]}\right] = E\left[\sum_{j=1}^{n}(n+1-j)P_{[j]}\right] = \sum_{j=1}^{n}(n+1-j)\mu_{[j]}$$

Variance of the total completion time:

$$\text{var}\left[\sum_{j=1}^{n} C_{[j]}\right] = \text{var}\left[\sum_{j=1}^{n}(n+1-j)P_{[j]}\right] = \sum_{j=1}^{n}(n+1-j)^2\sigma_{[j]}^2$$

Example 4.1. Application of the Expectation and Variance Expressions

Consider the following instance with four jobs. The mean and variance of the processing times for these jobs are given in Table 4.1.

There are 24 feasible sequences for scheduling the jobs on a single machine, and we analyze a few sequences that are candidates for a solution.

Note:
Here and henceforth, sequences that are optimal with respect to the expected value only will be indicated with an asterisk (*) and the other "good" and "reasonably optimal" sequences that are expectation-variance-efficient will be indicated with a double asterisk (**).

Sequence 2–4–1–3* has an expectation and variance of $E = 295$ and $V = 365$, respectively. This sequence is obtained by arranging the jobs in the shortest expected processing time (SEPT) order, and it is optimal with respect to the expected value of the total completion time. On the other hand, there are two sequences that are optimal with respect to the variance of the completion time. Sequences 4–1–2–3 and 4–2–1–3 can be obtained by arranging the jobs in the shortest processing time variance (SPTV) order, and their respective mean and variances are (320, 295) and (300, 295). However, sequence 4–2–1–3** is a "good" sequence with respect to both expected value and variance because its expected value is only slightly above that of sequence 2–4–1–3, whereas its variance is significantly lower than that of sequence 2–4–1–3. This indicates that by sacrificing a little on the expected value, significant gains can be achieved on the variance front. Such a schedule is attractive from a practical viewpoint because it induces lower variability in the system, thereby achieving "almost optimal" performance. The terms *good* and *almost optimal* are relative and can vary with the specific objectives of the scheduler. It depends on the scheduler's risk attitude or, in other words, his or her willingness to allow for higher or lower variance in the system.

4.2.2 Total Weighted Completion Time

In this formulation, every job $[j]$ has a corresponding weight $w_{[j]}$ associated with its completion time $C_{[j]}$. This weight could be considered an importance factor or, alternately, could represent either a holding cost per unit time or the value added to job $[j]$.

For a given schedule, the sum of weighted completion times of the jobs is given by

$$\sum_{j=1}^{n} w_{[j]} C_{[j]} = \sum_{j=1}^{n} w_{[j]} \left(\sum_{i=1}^{j} P_{[j]} \right)$$

By rearranging and summing up weights associated with each $P_{[j]}$, we have

$$\sum_{j=1}^{n} w_{[j]} C_{[j]} = \sum_{j=1}^{n} P_{[j]} G_{[j]}$$

where $G_{[j]} = \sum_{i=j}^{n} w_{[i]}$.

Expectation of the total weighted completion time:

$$E\left[\sum_{j=1}^{n} w_{[j]} C_{[j]} \right] = E\left[\sum_{j=1}^{n} P_{[j]} G_{[j]} \right] = \sum_{j=1}^{n} \mu_{[j]} G_{[j]}$$

Variance of the total weighted completion time:

$$\text{var}\left[\sum_{j=1}^{n} w_{[j]} C_{[j]} \right] = \text{var}\left[\sum_{j=1}^{n} P_{[j]} G_{[j]} \right] = \sum_{j=1}^{n} \sigma_{[j]}^2 G_{[j]}^2$$

To illustrate the application of these expressions, consider the data set for the four jobs shown in Table 4.1. The weights of the jobs are 5, 4, 10, and 5, respectively. If we designate by E the expected value of a sequence and by V its variance, we have

> Sequences*: 2–4–3–1 with $E = 2080$ and $V = 15515$
> 4–2–3–1 with $E = 2080$ and $V = 13170$
> Sequences**: 4–3–2–1 with $E = 2120$ and $V = 11690$
> 4–3–1–2 with $E = 2180$ and $V = 11555$

Note that sequence 4–2–3–1 has the same expected value as does sequence 2–4–3–1 but with a lower variance value. Sequence 4–3–2–1, on the other hand, has a slightly higher expected value but a much lower variance value than that of sequence 2–4–3–1 or sequence 4–2–3–1.

4.2.3 Total Weighted Discounted Completion Time

This is a more general cost function than the previous ones. The costs are now discounted at a rate of r, $0 < r < 1$, per unit time. That is, if job $[j]$ is not completed by time t, an additional cost $w_j r e^{-rt} dt$ is incurred over the period $[t, t + dt]$. If job $[j]$ is completed at time t, the total cost incurred over the period $[0, t]$ is $w_j(1 - e^{-rt})$. The value of r is usually close to zero (see Pinedo, 2002).

For a given schedule, the sum of the weighted discounted completion times of the jobs is

$$\sum_{j=1}^{n} w_{[j]}(1 - e^{-rC_{[j]}}) = \sum_{j=1}^{n} w_{[j]} - \sum_{j=1}^{n} w_{[j]} e^{-rC_{[j]}} = W - \sum_{j=1}^{n} w_{[j]} e^{-rC_{[j]}} \quad (4.1)$$

where $W = \sum_{j=1}^{n} w_j$.

Expectation of the total weighted discounted completion time:

Using Equation (4.1),

$$E\left[\sum_{j=1}^{n} w_{[j]}(1 - e^{-rC_{[j]}})\right] = W - \sum_{j=1}^{n} w_{[j]} \mu_{DC[j]} \quad (4.2)$$

where $\mu_{DC[j]} = E[e^{-rC_{[j]}}]$.

Variance of the total weighted discounted completion time:

$$\text{var}\left[\sum_{j=1}^{n} w_{[j]}(1 - e^{-rC_{[j]}})\right] = \text{var}\left[W - \sum_{j=1}^{n} w_{[j]} e^{-rC_{[j]}}\right]$$

$$= \text{var}\left[\sum_{j=1}^{n} w_{[j]} e^{-rC_{[j]}}\right], \text{ because var}[W] = 0 \quad (4.3)$$

Furthermore,

$$\text{var}\left[\sum_{j=1}^{n} w_{[j]} e^{-rC_{[j]}}\right] = \sum_{j=1}^{n} w_{[j]}^2 \text{var}[e^{-rC_{[j]}}] + 2 \sum_{i=1}^{n-1} \sum_{j=i+1}^{n} w_{[i]} w_{[j]} \text{cov}[e^{-rC_{[j]}}, e^{-rC_{[i]}}]$$

$$= \sum_{j=1}^{n} w_{[j]}^2 \sigma_{DC[j]}^2 + 2 \sum_{i=1}^{n-1} \sum_{j=i+1}^{n} w_{[i]} w_{[j]} \sigma_{DC[ij]} \quad (4.4)$$

where $\sigma_{DC[j]}^2 = \text{var}[e^{-rC_{[j]}}]$ and $\sigma_{DC[ij]} = \text{cov}[e^{-rC_{[i]}}, e^{-rC_{[j]}}]$.

Using Equation (4.4) in Equation (4.3), we have

$$\text{var}\left[\sum_{j=1}^{n} w_{[j]}(1 - e^{-rC_{[j]}})\right] = \sum_{j=1}^{n} w_{[j]}^2 \sigma_{DC[j]}^2 + 2\sum_{i=1}^{n-1}\sum_{j=i+1}^{n} w_{[i]}w_{[j]}\sigma_{DC[ij]} \quad (4.5)$$

4.2.3.1 Determination of $E[e^{-rC_{[j]}}]$

The moment-generating function of a random variable X is given by $M_X(t) = E(e^{tX})$, for any real value of t. In other words,

$$M_X(t) = \sum_{\text{all } i} e^{tx_i} p_X(x_i) \quad \text{(for discrete } X)$$

or

$$M_X(t) = \int_{-\infty}^{\infty} e^{tx} f_X(x)dx \quad \text{(for continuous } X)$$

Moreover, for sum of independent random variables, that is, $Y = X_1 + X_2 + \cdots + X_n$, the moment-generating function is given by

$$M_Y(t) = M_{X_1}(t)M_{X_2}(t)\cdots M_{X_n}(t)$$

$$= \prod_{j=1}^{n} M_{X_j}(t) \quad (4.6)$$

Hence $E[e^{-rC_{[j]}}]$ is the moment-generating function of the random variable $C_{[j]}$ with $t = -r$ and $0 < r < 1$. Additionally, $C_{[j]}$ is a random variable that is a sum of j independent random variables (processing times)

$$C_{[j]} = \sum_{i=1}^{j} P_{[i]}, \quad \text{where } 1 \le j \le n$$

The moment-generating function $E[e^{-rC_{[j]}}]$ thus can be determined using Equation (4.6) by the product of the moment-generating functions of the individual processing times.

$$M_{C_{[j]}}(-r) = \prod_{i=1}^{j} M_{P_{[i]}}(-r) \quad (4.7)$$

4.2.3.2 Determination of $\text{var}[e^{-rC_{[j]}}]$

$$\text{var}[e^{-rC_{[j]}}] = \sigma_{DC[j]}^2 = E[(e^{-rC_{[j]}})^2] - (E[e^{-rC_{[j]}}])^2 = E[e^{-2rC_{[j]}}] - (E[e^{-rC_{[j]}}])^2$$

4.2.3.3 Determination of $\text{cov}[e^{-rC_{[i]}}, e^{-rC_{[j]}}]$

$$\text{cov}[e^{-rC_{[i]}}, e^{-rC_{[j]}}] = E[e^{-rC_{[i]}} \cdot e^{-rC_{[j]}}] - (E[e^{-rC_{[i]}}] \cdot E[e^{-rC_{[j]}}])$$

Note that $E[e^{-rC_{[i]}} \cdot e^{-rC_{[j]}}] = E[e^{-r(C_{[i]}+C_{[j]})}]$; it is the moment-generating function of the random variable $C_{[i]}+C_{[j]}$, which, in turn, is the sum of j independent random variables (job processing times):

$$C_{[i]} + C_{[j]} = 2P_{[1]} + 2P_{[2]} + \cdots + 2P_{[i]} + P_{[i+1]} + \cdots + P_{[j]}, \quad \text{where } i < j$$

Using Equation (4.7),

$$M_{C_{[i]}+C_{[j]}}(-r)$$
$$= M_{P_{[1]}}(-2r)M_{P_{[2]}}(-2r) \cdots M_{P_{[i]}}(-2r)M_{P_{[i+1]}}(-r) \cdots M_{P_{[j]}}(-r)$$

The moment-generating functions can be determined after having known the probability density functions of the random job processing times. Standard formulas for moment-generating functions exist for all commonly used continuous distributions, such as exponential and normal, among others. For illustrative purposes, sample calculations are presented below for exponentially and normally distributed processing times.

Assuming that the job processing times are exponentially distributed with rates $\lambda_{[j]}$, the moment-generating function for the processing time of the job in the jth position of a sequence is given by

$$M_{P_{[j]}}(-r) = E[e^{-rP_{[j]}}] = \frac{\lambda_{[j]}}{\lambda_{[j]} + r}$$

From Equation (4.7),

$$M_{C_{[j]}}(-r) = E[e^{-rC_{[j]}}] = \left(\frac{\lambda_{[1]}}{\lambda_{[1]} + r}\right)\left(\frac{\lambda_{[2]}}{\lambda_{[2]} + r}\right) \cdots \left(\frac{\lambda_{[j]}}{\lambda_{[j]} + r}\right)$$

$$= \prod_{i=1}^{j}\left(\frac{\lambda_{[i]}}{\lambda_{[i]} + r}\right) \tag{4.8}$$

Applying Equation (4.8) in Equation (4.2), the expectation of the total weighted discounted completion time for exponentially distributed job processing times is given by

$$E\left[\sum w_{[j]}(1 - e^{-rC_{[j]}})\right] = W - \sum_{j=1}^{n}\left(w_{[j]}\prod_{i=1}^{j}\left(\frac{\lambda_{[i]}}{\lambda_{[i]} + r}\right)\right)$$

The variance can be evaluated as follows:

$$\text{var}[e^{-rC_{[j]}}] = \sigma_{DC[j]}^2 = E[(e^{-rC_{[j]}})^2] - (E[e^{-rC_{[j]}}])^2$$
$$= E[e^{-2rC_{[j]}}] - (E[e^{-rC_{[j]}}])^2$$
$$= \prod_{i=1}^{j}\left(\frac{\lambda_{[i]}}{\lambda_{[i]} + 2r}\right) - \prod_{i=1}^{j}\left(\frac{\lambda_{[i]}^2}{(\lambda_{[i]} + r)^2}\right) \tag{4.9}$$

$$\text{cov}[e^{-rC_{[i]}}, e^{-rC_{[j]}}] = E[e^{-rC_{[j]}} \cdot e^{-rC_{[i]}}] - (E[e^{-rC_{[i]}}] \cdot E[e^{-rC_{[j]}}])$$

$$E[e^{-rC_{[i]}} \cdot e^{-rC_{[j]}}] = \prod_{k=1}^{i} \left(\frac{\lambda_{[k]}}{\lambda_{[k]} + 2r} \right) \prod_{k=i+1}^{j} \left(\frac{\lambda_{[k]}}{\lambda_{[k]} + r} \right)$$

$$\sigma_{DC[ij]} = \prod_{k=1}^{i} \left(\frac{\lambda_{[k]}}{\lambda_{[k]} + 2r} \right) \prod_{k=i+1}^{j} \left(\frac{\lambda_{[k]}}{\lambda_{[k]} + r} \right)$$

$$- \prod_{k=1}^{i} \left(\frac{\lambda_{[k]}}{\lambda_{[k]} + r} \right) \prod_{k=1}^{j} \left(\frac{\lambda_{[k]}}{\lambda_{[k]} + r} \right) \tag{4.10}$$

Expressions (4.9) and (4.10) can be substituted in (4.5) to obtain the variance of the total weighted discounted completion time, $\text{var}[\sum w_{[j]}(1 - e^{rC_{[j]}})]$, for the exponentially distributed job processing times.

We, next, illustrate the application of the above expressions to the data shown in Table 4.1. Assume the variances to be that of exponentially distributed job processing times. Let the discount rate, $r = 0.05$.

Example 4.2. Application of the Expectation and Variance Expressions

Developed Earlier for Exponentially Distributed Job Processing Times

Sequences*:   3–4–1–2 with $E = 7.0093$ and $V = 11.6663$
              4–3–1–2 with $E = 7.0093$ and $V = 9.2676$
Sequences **: 4–3–2–1 with $E = 7.1092$ and $V = 9.1870$
              4–1–2–3 with $E = 8.0510$ and $V = 8.8156$

Note that the sequences that minimize the expected weighted discounted completion time are different from those that have lower variance of the total weighted discounted completion time.

Assuming that the job processing times are normally distributed with means $\mu_{[j]}$ and variances $\sigma_{[j]}^2$, the moment-generating function for the processing time of the job in the jth position of a sequence is given by

$$M_{P_{[j]}}(-r) = E[e^{-rp_{[j]}}] = e^{(-r\mu_{[j]} + \sigma_{[j]}^2 r^2/2)}$$

From Equation (4.7),

$$M_{C_{[j]}}(-r) = E[e^{-rC_{[j]}}]$$

$$= e^{(-r\mu_{[1]} + \sigma_{[1]}^2 r^2/2)} e^{(-r\mu_{[2]} + \sigma_{[2]}^2 r^2/2)} \cdots e^{(-r\mu_{[j]} + \sigma_{[j]}^2 r^2/2)}$$

$$= e^{(-r(\mu_{[1]} + \mu_{[2]} + \cdots + \mu_{[j]}) + (\sigma_{[1]}^2 + \sigma_{[2]}^2 + \cdots + \sigma_{[j]}^2)r^2/2)}$$

Denoting $\mu_{C_{[j]}} = E[C_{[j]}] = \sum_{k=1}^{j} \mu_{[k]}$ and $\sigma_{C_{[j]}}^2 = \text{var}[C_{[j]}] = \sum_{k=1}^{j} \sigma_{[k]}^2$, we have

$$\mu_{DC[j]} = E[e^{-rC_{[j]}}] = e^{(-r\mu_{C_{[j]}} + \sigma_{C_{[j]}}^2 r^2/2)} \tag{4.11}$$

Using Equation (4.11) in Equation (4.2), the expectation of the total weighted discounted completion time for normally distributed job processing times is given by

$$E\left[\sum w_{[j]}(1 - e^{-rC_{[j]}})\right] = W - \sum_{j=1}^{n}(w_{[j]}e^{(-r\mu_{C_{[j]}} + \sigma_{C_{[j]}}^2 r^2/2)})$$

Meanwhile,

$$\text{var}[e^{-rC_{[j]}}] = \sigma_{DC[j]}^2 = E[e^{-2rC_{[j]}}] - (E[e^{-rC_{[j]}}])^2$$
$$= e^{(-2r\mu_{C_{[j]}} + 2\sigma_{C_{[j]}}^2 r^2)} - e^{(-2r\mu_{C_{[j]}} + \sigma_{C_{[j]}}^2 r^2)} \tag{4.12}$$

$$\text{cov}[e^{-rC_{[i]}}, e^{-rC_{[j]}}] = E[e^{-rC_{[i]}} \cdot e^{-rC_{[j]}}] - (E[e^{-rC_{[i]}}] \cdot E[e^{-rC_{[j]}}])$$
$$= E[e^{-r(C_{[i]} + C_{[j]})}] - (E[e^{-rC_{[i]}}] \cdot E[e^{-rC_{[j]}}])$$

Denoting $\mu_{C_{[i]} + C_{[j]}} = E[C_{[i]} + C_{[j]}] = 2\sum_{k=1}^{i} \mu_{[k]} + \sum_{k=i+1}^{j} \mu_{[k]}$ and

$$\sigma_{C_{[i]} + C_{[j]}}^2 = \text{var}[C_{[i]} + C_{[j]}] = 4\sum_{k=1}^{i} \sigma_{[k]}^2 + \sum_{k=i+1}^{j} \sigma_{[k]}^2$$

we get

$$\sigma_{DC[ij]} = e^{(-r\mu_{C_{[i]} + C_{[j]}} + \sigma_{C_{[i]} + C_{[j]}}^2 r^2/2)} - e^{(-r\mu_{C_{[i]}} + \sigma_{C_{[i]}}^2 r^2/2)} \cdot e^{(-r\mu_{C_{[j]}} + \sigma_{C_{[j]}}^2 r^2/2)} \tag{4.13}$$

Equations (4.12) and (4.13) can be substituted in Equation (4.5) to obtain the variance of the total weighted discounted completion time, $\text{var}[\sum w_{[j]}(1 - e^{-rC_{[j]}})]$, for normally distributed job processing times.

A similar approach could be adopted to compute the expectation and variance of the total weighted discounted completion time for jobs with other standard distributions of processing times.

As an illustration of the use of the preceding expressions, consider the data depicted in Table 4.1. Let the means and variances be those of the normally distributed job processing times and the discount rate $r = 0.05$.

Sequence*:   2–4–1–3 with $E = 21.8782$ and $V = 0.1774$
Sequences**: 2–4–3–1 with $E = 21.9013$ and $V = 0.1732$
              4–2–1–3 with $E = 22.0441$ and $V = 0.0586$

4–2–3–1 with $E = 22.0673$ and $V = 0.0565$
4–1–2–3 with $E = 22.2925$ and $V = 0.0396$
4–1–3–2 with $E = 22.3365$ and $V = 0.0368$
4–3–2–1 with $E = 22.3855$ and $V = 0.0344$
4–3–1–2 with $E = 22.3981$ and $V = 0.0336$
1–4–3–2 with $E = 23.0884$ and $V = 0.0324$
1–3–4–2 with $E = 23.2269$ and $V = 0.0232$
3–2–4–1 with $E = 23.3816$ and $V = 0.0201$
3–4–2–1 with $E = 23.3901$ and $V = 0.0192$
3–2–1–4 with $E = 23.3960$ and $V = 0.0191$
3–4–1–2 with $E = 23.4028$ and $V = 0.0184$
3–1–2–4 with $E = 23.4400$ and $V = 0.0162$
3–1–4–2 with $E = 23.4411$ and $V = 0.0161$

Note that there are quite a few expectation-variance-efficient sequences in this case. The variance of the total weighted discounted completion times for sequence 3–1–4–2 is considerably smaller than that for sequence 2–4–1–3 with a slight increment in its expected value over that for sequence 2–4–1–3.

4.3 Due-Date-Based Objectives
4.3.1 Total Tardiness

Recall that tardiness $T_{[j]}$ for a job is given by

$$T_{[j]} = \max(L_{[j]}, 0) = \max(C_{[j]} - d_{[j]}, 0)$$

The expectation of total tardiness is given by

$$E\left[\sum_{j=1}^{n} T_{[j]}\right] = \sum_{j=1}^{n} E[T_{[j]}] = \sum_{j=1}^{n} \mu_{T_{[j]}}, \quad \text{where } \mu_{T_{[j]}} = E[T_{[j]}] \quad (4.14)$$

The variance of total tardiness is similarly given by

$$\text{var}\left[\sum_{j=1}^{n} T_{[j]}\right] = \sum_{j=1}^{n} \text{var}[T_{[j]}] + 2\sum_{i=1}^{n-1} \sum_{j=i+1}^{n} \text{cov}[T_{[i]}, T_{[j]}]$$

$$= \sum_{j=1}^{n} \sigma^2_{T_{[j]}} + 2\sum_{i=1}^{n-1} \sum_{j=i+1}^{n} \sigma_{T[ij]} \quad (4.15)$$

where $\sigma^2_{T_{[j]}} = \text{var}[T_{[j]}]$, and $\sigma_{T[ij]} = \text{cov}[T_{[i]}, T_{[j]}]$.

The complexity of computing $\mu_{T_{[j]}}$ and $\sigma^2_{T_{[j]}}$ in Equations (4.14) and (4.15) arises from the fact that we need to compute the expectation and variance of a maximum function, such as

$$\mu_{T[j]} = E[\max(L_{[j]}, 0)] = E[\max(C_{[j]} - d_{[j]}, 0)]$$

and

$$\sigma_{T[j]}^2 = \text{var}[\max(L_{[j]}, 0)] = \text{var}[\max(C_{[j]} - d_{[j]}, 0)]$$

There are no standard expressions available for the mean and variance of a maximum operator in mathematical theory. However, in a pioneering work, Clark (1961) developed a method to recursively estimate the expectation and variance of the greatest of a finite set of random variables that are normally distributed. Accurate results can be obtained for maximum functions with two arguments, whereas for higher numbers of arguments, close-to-accurate approximations can be obtained. Hence, assuming normally distributed processing times for all the jobs, Clark's method could be applied to find the desired expectation and variance expressions of $\sum T_j$.

From the definition of tardiness, it is also evident that the tardiness random variables ($T_{[j]}$) are not independent (because they are correlated by their individual completion times), and hence it is necessary to compute the covariance between every pairs of $T_{[j]}$ values.

The covariance between $T_{[i]}$ and $T_{[j]}$ is given by

$$\text{cov}[T_{[i]}, T_{[j]}] = \sigma_{T[ij]} = E[T_{[i]} \cdot T_{[j]}] - E[T_{[i]}]E[T_{[j]}]$$

$$= E[T_{[i]} \cdot T_{[j]}] - \mu_{T[i]}\mu_{T[j]} \tag{4.16}$$

Before we proceed to evaluate $\mu_{T[j]}, \sigma_{T[j]}^2$, and $E[T_{[i]}T_{[j]}]$, we first develop certain basic expressions leading from our assumption of a normal distribution for job processing times.

$C_{[i]}$ and $C_{[j]}$ are linear sums of the job processing times that are assumed to be normally distributed. By the reproductive property of normal random variables, $C_{[i]}$ and $C_{[j]}$ are also normally distributed.

Hence, for example, $\mu_{c[j]} = E[C_{[j]}] = \sum_{k=1}^{j} \mu_{[k]}$ and $\sigma_{c[j]}^2 = \text{var}[C_{[j]}] = \sum_{k=1}^{j} \sigma_{[k]}^2$. The probability density function for $C_{[j]}$ is $f(x) = \frac{1}{\sigma_{C[j]}\sqrt{2\pi}}\exp\left(\frac{-(x-\mu_{C[j]})^2}{2\sigma_{C[j]}^2}\right)$.

Recall that

$$T_{[j]} = \max(L_{[j]}, 0) = \max(C_{[j]} - d_{[j]}, 0)$$

The second argument inside the maximum function, viz., 0, can be assumed to be a random variable with mean $\mu = 0$ and variance $\sigma^2 = 0$. The mean and variance of the first argument, $C_{[j]} - d_{[j]}$, are given below. Hence Clark's method could be applied to accurately determine the expectation and variance of $\max(C_{[j]} - d_{[j]}, 0)$.

$$\mu_{[J]} \equiv E[C_{[j]} - d_{[j]}] = E[C_{[j]}] - d_{[j]} = \sum_{k=1}^{j} \mu_{[k]} - d_{[j]}$$

$$\sigma_{[J]}^2 \equiv \text{var}[C_{[j]} - d_{[j]}] = \text{var}[C_{[j]}] = \sum_{k=1}^{j} \sigma_{[k]}^2$$

The coefficient of linear correlation between the two arguments (0 and $T_{[j]}$) inside the maximum function is zero. $\mu_{T_{[j]}}$ thus is given by the first moment $v_{1[j]}$ of the random variable $\max(C_{[j]} - d_{[j]}, 0)$:

$$v_{1[j]} = \mu_{T_{[j]}} = \mu_{[J]} \Phi(\alpha_{[j]}) + a_{[j]} \varphi(\alpha_{[j]}) \tag{4.17}$$

The second moment is given by

$$v_{2[j]} = (\mu_{[J]}^2 + \sigma_{[J]}^2) \Phi(\alpha_{[j]}) + \mu_{[J]} a_{[j]} \varphi(\alpha_{[j]})$$

$$\sigma_{T_{[j]}}^2 = v_{2[j]} - v_{1[j]}^2$$

$$= [(\mu_{[J]}^2 + \sigma_{[J]}^2) \Phi(\alpha_{[j]}) + \mu_{[J]} a_{[j]} \varphi(\alpha_{[j]})] - [\mu_{[J]} \Phi(\alpha_{[j]}) + a_{[j]} \varphi(\alpha_{[j]})]^2 \tag{4.18}$$

where $a_{[j]}^2 = \sigma_{[J]}^2, \sigma_{[j]} = \mu_{[J]}/a_{[j]}, \varphi(z) = \frac{1}{\sqrt{2\pi}} \exp(-z^2/2)$, and $\Phi(x) = \int_{-\infty}^{x} \varphi(t) dt$.

Now, the only remaining part of Equations (4.14) and (4.15) that needs to be evaluated is $E[T_{[i]} T_{[j]}]$. To that end, let X and Y be two random variables such that X is the sum of the processing times of the first i jobs in the sequence and Y is the sum of the processing times of the jobs from position $(i + 1)$ to j in the given sequence. In other words, X and Y are $C_{[j]}$ and $C_{[j]} - C_{[i]}$ respectively. It is assumed that job i precedes job j in the sequence (see Figure 4.1).

Similar to the earlier discussion, X and Y are linear sums of the job processing times that are assumed to be normally distributed. By the reproductive property of normal random variables, X and Y are also normally distributed.

Hence the expectation of X is given by

$$E[X] = \mu_X = E[C_{[i]}] = \sum_{k=1}^{i} \mu_{[k]} \quad \text{and} \quad \text{var}[X] = \sigma_X^2 = \text{var}[C_{[i]}] = \sum_{k=1}^{i} \sigma_{[k]}^2$$

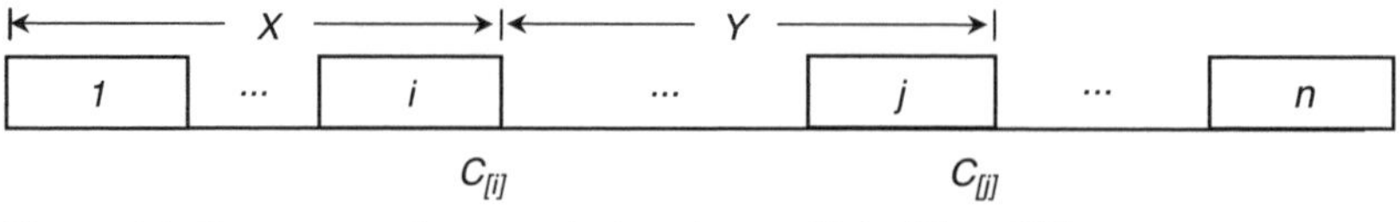

Figure 4.1. Representative completion times of jobs $[i]$ and $[j]$.

The probability density function for X is

$$f_X(x) = \frac{1}{\sigma_X\sqrt{2\pi}}\exp\left(\frac{-(x-\mu_X)^2}{2\sigma_X^2}\right)$$

Similarly, the expectation of Y is given by

$$E[Y] = \mu_Y = E[C_{[j]} - C_{[i]}] = \sum_{k=i+1}^{j} \mu_{[k]} \quad \text{and}$$

$$\text{var}[Y] = \sigma_Y^2 = \text{var}[C_{[j]} - C_{[i]}] = \sum_{k=i+1}^{j} \sigma_{[k]}^2$$

The probability density function for Y is

$$f_Y(y) = \frac{1}{\sigma_Y\sqrt{2\pi}}\exp\left(\frac{-(y-\mu_Y)^2}{2\sigma_Y^2}\right)$$

By definition, X and Y are independent random variables, and therefore, the joint probability density function for X and Y $f_{XY}(x,y) = f_X(x) \cdot f_Y(y)$ is

$$f_{XY}(x,y) = \frac{1}{\sigma_X\sqrt{2\pi}}\exp\left(\frac{-(x-\mu_X)^2}{2\sigma_X^2}\right) \cdot \frac{1}{\sigma_Y\sqrt{2\pi}}\exp\left(\frac{-(y-\mu_Y)^2}{2\sigma_Y^2}\right)$$

$$= \frac{1}{2\pi\sigma_X\sigma_Y}\exp\left(-\frac{1}{2\sigma_X^2\sigma_Y^2}(\sigma_Y^2(x-\mu_X)^2 + \sigma_X^2(y-\mu_Y)^2)\right)$$

Now $E[T_{[i]}T_{[j]}]$ can be evaluated as follows:

$$T_{[i]}T_{[j]} = \max(C_{[i]} - d_{[i]},0) \cdot \max(C_{[j]} - d_{[j]},0)$$

$$= \max(C_{[i]} - d_{[i]},0) \cdot \max((C_{[j]} - C_{[i]}) + C_{[i]} - d_{[j]},0)$$

$$= \max(X - d_{[i]},0) \cdot \max(Y + X - d_{[j]},0)$$

Note that $T_{[i]}T_{[j]}$ is a function of two independent random variables X and Y, and we know that

$$E[g(X,Y)] = \iint g(x,y)f_{XY}(x,y)dxdy$$

Therefore,

$$E[T_{[i]}T_{[j]}] = \int_{y=-\infty}^{\infty}\int_{x=-\infty}^{\infty} \max(x - d_{[i]},0) \cdot \max(y + x - d_{[j]},0)f_{XY}(x,y)dxdy$$

$$(4.19)$$

Since the integrand will be zero for any value of x and y such that $x < d_{[j]}$ or $y < d_{[j]} - x$, the expression can be simplified to

$E[T_{[i]}T_{[j]}]$

$$= \int_{x=d_{[i]}}^{\infty} \int_{y=d_{[j]}-x}^{\infty} (x - d_{[i]}) \cdot (y + x - d_{[j]}) f_{XY}(x,y) dy dx$$

$$= \int_{x=d_{[i]}}^{\infty} \int_{y=d_{[j]}-x}^{\infty} (x - d_{[i]}) \cdot [(y - \mu_Y) + (x - d_{[j]} + \mu_Y)] f_X(x) \cdot f_Y(y) dy dx$$

$$= \int_{x=d_{[i]}}^{\infty} (x - d_{[i]}) \cdot (x - d_{[j]} + \mu_Y) \frac{1}{\sqrt{2\pi}\sigma_X} \exp\left(-\frac{(x - \mu_X)^2}{2\sigma_X^2}\right)$$

$$\times \int_{y=d_{[j]}-x}^{\infty} \frac{1}{\sqrt{2\pi}\sigma_Y} \exp\left(-\frac{(y - \mu_Y)^2}{2\sigma_Y^2}\right) dy dx$$

$$+ \int_{x=d_{[i]}}^{\infty} (x - d_{[i]}) \cdot \frac{1}{\sqrt{2\pi}\sigma_X} \exp\left(-\frac{(x - \mu_X)^2}{2\sigma_X^2}\right)$$

$$\cdot \int_{y=d_{[j]}-x}^{\infty} (y - \mu_Y) \frac{1}{\sqrt{2\pi}\sigma_Y} \exp\left(-\frac{(y - \mu_Y)^2}{2\sigma_Y^2}\right) dy dx$$

$$= \int_{x=d_{[i]}}^{\infty} (x - d_{[i]}) \cdot (x - d_{[j]} + \mu_Y) \frac{1}{\sigma_X} \varphi\left(\frac{x - \mu_X}{\sigma_X}\right) \left[1 - \Phi\left(\frac{d_{[j]} - x - \mu_Y}{\sigma_Y}\right)\right] dx$$

$$+ \int_{x=d_{[i]}}^{\infty} (x - d_{[i]}) \cdot \frac{1}{\sigma_X} \varphi\left(\frac{x - \mu_X}{\sigma_X}\right) \cdot \left[-\frac{\sigma_Y}{\sqrt{2\pi}} \exp\left(-\frac{(y - \mu_Y)^2}{2\sigma_Y^2}\right) \Big|_{d_{[j]}-x}^{\infty}\right] dx$$

$$= \int_{x=d_{[i]}}^{\infty} (x - d_{[i]}) \cdot (x - d_{[j]} + \mu_Y) \frac{1}{\sigma_X} \varphi\left(\frac{x - \mu_X}{\sigma_X}\right) \Phi\left(\frac{x - d_{[j]} + \mu_Y}{\sigma_Y}\right) dx$$

$$+ \int_{x=d_{[i]}}^{\infty} (x - d_{[i]}) \cdot \frac{1}{\sigma_X} \varphi\left(\frac{x - \mu_X}{\sigma_X}\right) \cdot \sigma_Y \varphi\left(\frac{x - d_{[j]} + \mu_Y}{\sigma_Y}\right) dx \qquad (4.20)$$

Equation (4.20) can be evaluated numerically. Note that we can first apply the substitution $s = (x - d_{[i]})/(1 + x - d_{[i]})$ so that the region of integration becomes a finite interval.

Next, we provide a numerical example to illustrate the use of the preceding expressions. Consider the problem presented in Table 4.1 and the sequence 4–2–3–1. We start with the calculation of $\mu_{T[j]}$ and $\sigma_{T[j]}^2$. Taking $T_{[3]}$ for example, we have

$$\mu_{[J]} = E[C_{[3]} - d_{[3]}] = \sum_{k=1}^{3} \mu_{[k]} - d_{[3]} = (25 + 20 + 60) - 90 = 15$$

$$\sigma_{[J]}^2 = \text{var}[C_{[3]} - d_{[3]}] = \sum_{k=1}^{3} \sigma_{[k]}^2 = 5 + 15 + 20 = 40$$

By Equations (4.17) and (4.18), we can obtain $\mu_{T[j]}$ and $\sigma_{T[j]}^2$.

In the calculation of $\sigma_{T[ij]}$, consider $\sigma_{T[23]}$ as an example. Then

$$\mu_X = E[C_{[2]}] = 25 + 20 = 35 \quad \sigma_X^2 = \text{var}[C_{[i]}] = 5 + 15 = 20$$

$$\mu_Y = E[C_{[3]} - C_{[2]}] = 60$$

$$\sigma_Y^2 = \text{var}[C_{[3]} - C_{[2]}] = 20$$

By Equation (4.20) and, subsequently, Equation (4.16), we obtain $\sigma_{T[23]}$.
Furthermore, by Equations (4.14) and (4.15), we have

$$E\left[\sum_{j=1}^{n} T_{[j]}\right] \approx 40.019$$

$$\text{var}\left[\sum_{j=1}^{n} T_{[j]}\right] \approx 173.596$$

The foregoing process is repeated for all possible sequences, and the results
are shown in Figure 4.2.

Note that

Sequence*: 4–2–3–1 with $E = 40.0$ and $V = 173.6$
Sequence**: 4–2–1–3 with $E = 55.0$ and $V = 55.0$

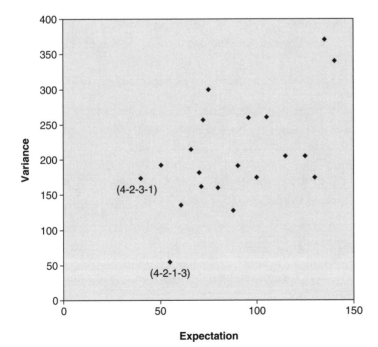

Figure 4.2. Numerical results for the total tardiness.

Although the sequence 4–2–3–1 gives the optimal value of expected total tardiness, sequence 4–2–1–3 results in a variance that is less than one-third that for sequence 4–2–3–1 with a relatively small increase in expectation.

4.3.2 Total Weighted Tardiness

The expressions for the expectation and variance for the total weighted tardiness measure are as follows:

$$E\left[\sum_{j=1}^{n} w_{[j]} T_{[j]}\right] = \sum_{j=1}^{n} E\left[w_{[j]} T_{[j]}\right] = \sum_{j=1}^{n} w_{[j]} \mu_{T[j]} \qquad (4.21)$$

where $\mu_{T[j]} = E[T_{[j]}]$.

$$\text{var}\left[\sum_{j=1}^{n} w_{[j]} T_{[j]}\right] = \sum_{j=1}^{n} w_{[j]}^2 \text{var}[T_{[j]}] + 2\sum_{i=1}^{n-1}\sum_{j=i+1}^{n} w_{[i]} w_{[j]} \text{cov}[T_{[i]}, T_{[j]}]$$

$$= \sum_{j=1}^{n} w_{[j]}^2 \sigma_{T[j]}^2 + 2\sum_{i=1}^{n-1}\sum_{j=i+1}^{n} w_{[i]} w_{[j]} \sigma_{T[ij]} \qquad (4.22)$$

where $\sigma_{T[j]}^2 = \text{var}[T_{[j]}]$, and $\sigma_{T[ij]} = \text{cov}[T_{[i]}, T_{[j]}]$.

The determination of $E\left[\sum_{j=1}^{n} w_{[j]} T_{[j]}\right]$ and $\text{var}\left[\sum_{j=1}^{n} w_{[j]} T_{[j]}\right]$ clearly follows from the development in Section 4.3.1. The expressions for $\mu_{T[j]}$ and $\sigma_{T[j]}^2$ are developed using Clark's equations, whereas the $\sigma_{T[ij]}$ values are calculated by numerical integration.

Applying the data in Table 4.1 to the total weighted tardiness case, we have

Sequence*:   4–3–2–1 with $E = 229.2$ and $V = 4498.2$
Sequences**: 4–3–1–2 with $E = 293.0$ and $V = 3663.3$
              2–3–1–4 with $E = 440.2$ and $V = 3385.4$

Sequence 4–3–1–2 results in some improvement in variance with not much increment in the expected value over sequence 4–3–2–1.

4.3.3 Total Number of Tardy Jobs

The fact that a job is tardy is captured by defining a unit penalty function for job j as follows:

$$U_{[j]} = \begin{cases} 1, & \text{if } C_{[j]} > d_{[j]} \\ 0, & \text{if } C_{[j]} \le d_{[j]} \end{cases} \qquad (4.23)$$

The expectation and variance of the total number of tardy jobs for a given sequence are given as follows:

$$E\left[\sum_{j=1}^{n} U_{[j]}\right] = \sum_{j=1}^{n} E[U_{[j]}] = \sum_{j=1}^{n} \mu_{U[j]} \tag{4.24}$$

$$\mathrm{var}\left[\sum_{j=1}^{n} U_{[j]}\right] = \sum_{j=1}^{n} \mathrm{var}[U_{[j]}] + 2\sum_{i=1}^{n}\sum_{j=i+1}^{n} \mathrm{cov}[U_{[i]}, U_{[j]}]$$

$$= \sum_{j=1}^{n} \sigma^2_{U[j]} + 2\sum_{i=1}^{n}\sum_{j=i+1}^{n} \sigma_{U[ij]} \tag{4.25}$$

where $\sigma_{U[ij]} = E[U_{[i]}U_{[j]}] - \mu_{U[i]} \cdot \mu_{U[j]}$.

Note that the random variable $U_{[j]}$ involves two outcomes, one for a job being early and another for it being late, and hence can be modeled using the Bernoulli distribution. Therefore,

$$\mu_{U[j]} = (1 \cdot \Pr[C_{[j]} > d_{[j]}] + 0 \cdot \Pr[C_{[j]} \le d_{[j]}])$$

$$= 1 - \Pr[C_{[j]} \le d_{[j]}] = 1 - F_{C[j]}(d_{[j]}) \tag{4.26}$$

$$\sigma^2_{U[j]} = \mu_{U[j]}(1 - \mu_{U[j]}) \tag{4.27}$$

Substituting Equations (4.24) and (4.25) in Equations (4.26) and (4.27), we have

$$E\left[\sum_{j=1}^{n} U_{[j]}\right] = \sum_{j=1}^{n}(1 - F_{C[j]}(d_{[j]}))$$

and

$$\mathrm{var}\left[\sum_{j=1}^{n} U_{[j]}\right] = \sum_{j=1}^{n}[(1 - F_{C[j]}(d_{[j]})) \cdot F_{C[j]}(d_{[j]})]$$

$$+ 2\sum_{i=1}^{n-1}\sum_{j=i+1}^{n}[E[U_{[i]}U_{[j]}] - (1 - F_{C[i]}(d_{[i]})) \cdot (1 - F_{C[j]}(d_{[j]}))]$$

Similar to the calculation of $E[T_{[i]}T_{[j]}]$, we have

$$E[U_{[i]}U_{[j]}] = \int_{x=d_{[i]}}^{\infty} \int_{y=d_{[j]}-x}^{\infty} f_X(x)f_Y(y)dydx$$

As an illustration for determining the numeric values, we consider normally distributed job processing times. Since $C_{[i]}$ and $C_{[j]}$ are linear sums of job processing times that are normally distributed, they are normally distributed themselves. Hence

$$F_{C[j]}(d_{[j]}) = \Phi\left(\frac{d_{[j]} - \mu_{C[j]}}{\sigma_{C[j]}}\right)$$

Furthermore,

$$
\begin{aligned}
E[U_{[i]}U_{[j]}] &= \int_{x=d_{[i]}}^{\infty} \int_{y=d_{[j]}-x}^{\infty} f_X(x)f_Y(y)dydx \\
&= \int_{x=d_{[i]}}^{\infty} \frac{1}{\sqrt{2\pi}\sigma_X} \exp\left(-\frac{(x-\mu_X)^2}{2\sigma_X^2}\right) \\
&\quad \times \int_{y=d_{[j]}-x}^{\infty} \frac{1}{\sqrt{2\pi}\sigma_Y} \exp\left(-\frac{(y-\mu_Y)^2}{2\sigma_Y^2}\right) dydx \\
&= \int_{x=d_{[i]}}^{\infty} \frac{1}{\sigma_X} \varphi\left(\frac{x-\mu_X}{\sigma_X}\right) \Phi\left(\frac{x-d_{[j]}+\mu_Y}{\sigma_Y}\right) dx \quad (4.28)
\end{aligned}
$$

We can use numerical integration to calculate $E[U_{[i]}U_{[j]}]$.

Again, considering the data given in Table 4.1 and assuming the means and variances to be those of the underlying normal distribution, the evaluation of various sequences for this case is as follows:

Sequence*:   4–2–1–3 with $E = 1.000$ and $V = 1.6 \times 10^{-9}$
Sequence**: 4–2–1–3 with $E = 1.000$ and $V = 1.6 \times 10^{-9}$

Note that for this sample data set, the sequence that minimizes the expected value also minimizes the variance (for the level of significance shown); that is, we only have one EV-efficient sequence.

4.3.4 Total Weighted Number of Tardy Jobs

The expectation and variance for the total weighted number of tardy jobs for a given sequence can be expressed as follows:

$$E\left[\sum_{j=1}^{n} w_{[j]}U_{[j]}\right] = \sum_{j=1}^{n} w_{[j]}E[U_{[j]}] = \sum_{j=1}^{n} w_{[j]}\mu_{U[j]} \quad (4.29)$$

$$
\begin{aligned}
\text{var}\left[\sum_{j=1}^{n} w_{[i]}U_{[j]}\right] &= \sum_{i=1}^{n} w_{[j]}^2 \, \text{var}[U_{[j]}] + 2\sum_{i=1}^{n-1}\sum_{j=i+1}^{n} w_{[i]}w_{[j]}\text{cov}[U_{[i]}, U_{[j]}] \\
&= \sum_{j=1}^{n} w_{[j]}^2 \sigma_{U[j]}^2 + 2\sum_{i=1}^{n-1}\sum_{j=i+1}^{n} w_{[i]}w_{[j]}\sigma_{U[ij]} \quad (4.30)
\end{aligned}
$$

Similar to the analysis presented in the preceding section for the performance measure of total number of tardy jobs, we can use, in these expressions, the expressions for $\mu_{U[j]}$, $\sigma^2_{U[j]}$, and $\sigma_{U[ij]}$ presented there to evaluate the expectation and variance for the total weighted number of tardy jobs.

Again, assuming the means and variances given in Table 4.1 to be those of the underlying normal distributions of processing times, the evaluations of various sequences for this case are as follows:

Sequence*:   2–3–1–4 with $E = 7.955$ and $V = 12.859$
Sequences**: 3–1–2–4 with $E = 9.002$ and $V = 0.009$
             4–2–1–3 with $E = 10.000$ and $V = 4.1 \times 10^{-8}$

Both sequences 3–1–2–4 and 4–2–1–3 result in relatively large improvements in variance with small increments in the expected values.

4.3.5 Mean Lateness

The mean lateness for a given schedule is given by

$$\overline{L} = \frac{1}{n}\left[\sum_{j=1}^{n} L_{[j]}\right] = \frac{1}{n}\left[\sum_{j=1}^{n}(C_{[j]} - d_{[j]})\right]$$

$$= \frac{1}{n}\left[\sum_{j=1}^{n} C_{[j]} - \sum_{j=1}^{n} d_{[j]}\right]$$

Accordingly, the expectation of $\overline{L}$ is given by

$$E[\overline{L}] = E\left[\frac{1}{n}\left(\sum_{j=1}^{n} C_{[j]} - \sum_{j=1}^{n} d_{[j]}\right)\right] = \frac{1}{n}\left(E\left[\sum_{j=1}^{n} C_{[j]}\right] - \sum_{j=1}^{n} d_{[j]}\right)$$

However, $E\left[\sum_{j=1}^{n} C_{[j]}\right]$ has already been shown to equal $\sum_{j=1}^{n}(n+1-j)\mu_{[j]}$ (see Section 4.2.1). Therefore,

$$E[\overline{L}] = \frac{1}{n}\left[\sum_{j=1}^{n}(n+1-j)\mu_{[j]} - \sum_{j=1}^{n} d_{[j]}\right]$$

Also,

$$\text{var}[\overline{L}] = \text{var}\left[\frac{1}{n}\left(\sum_{j=1}^{n} C_{[j]} - \sum_{j=1}^{n} d_{[j]}\right)\right] = \frac{1}{n^2}\left(\text{var}\left[\sum_{j=1}^{n} C_{[j]}\right]\right)$$

$$= \frac{1}{n^2}\left[\sum_{j=1}^{n}(n+1-j)^2\mu^2_{[j]}\right] \quad \text{(see Section 4.2.1)}$$

Application of these expressions to the data set in Table 4.1 results in

Sequence*: 2–4–1–3 with $E = -13.750$ and $V = 22.813$
Sequence**: 4–2–1–3 with $E = -12.500$ and $V = 18.438$

Note that for this sample data set, schedule 2–4–1–3 is optimal with respect to the expected value, and schedule 4–2–1–3 is optimal with respect to variance.

4.3.6 Maximum Lateness

Next, we consider the criterion of maximum lateness. Let

$$L_{\max} = \max(L_{[1]}, L_{[2]}, \ldots, L_{[n-1]}, L_{[n]}) \tag{4.31}$$

$L_{\max}$ is the maximum of n random variables. Clark's equations can be recursively applied to approximately determine the expectation and variance under the assumption that the lateness variables are mutually related by the multivariate normal distribution.

Note that the $L_{\max}$ function can be expressed as follows:

$$L_{\max} = \max(\max(L_{[1]}, L_{[2]}, \ldots, L_{[n-1]}), L_{[n]})$$

$$\vdots$$

$$= \max(\max(\ldots \max(\max(L_{[1]}, L_{[2]}), L_{[3]}), \ldots, L_{[n-1]}), L_{[n]})$$

The mean and variance of the lateness of a job j, that is $L_{[j]}$, are given by

$$\mu_{L[J]} = E[C_{[j]} - d_{[j]}] = E[C_{[j]}] - d_{[j]} = \sum_{k=1}^{j} \mu_{[k]} - d_{[j]}$$

$$\sigma^2_{L[J]} = \text{var}[C_{[j]} - d_{[j]}] = \text{var}[C_{[j]}] = \sum_{k=1}^{j} \sigma^2_{[k]}$$

Given that the processing times of the jobs are normally distributed, the random variable $L_{[J]}$ is also normally distributed with mean $\mu_{L[J]}$ and variance $\sigma^2_{L[J]}$.

For ease of subsequent development, let $r(L_{[i]}, L_{[j]}) = $ coefficient of linear correlation $(\rho_{[ij]})$,

$$a^2_{[ij]} = \sigma^2_{L[i]} + \sigma^2_{L[j]} - 2\sigma_{L[i]}\sigma_{L[j]}\rho_{[ij]}$$

and

$$\alpha_{[ij]} = \frac{\mu_{L[i]} - \mu_{L[j]}}{a_{[ij]}}$$

Let $L_{[i]} = C_{[i]} - d_{[i]} = X - d_{[i]}$ and $L_{[j]} = C_{[j]} - d_{[j]} = X + Y - d_{[j]}$, where X and Y are as defined in Section 4.3.1. Then we have

$$E[L_{[i]}L_{[j]}] = \int_{x=-\infty}^{\infty} \int_{y=-\infty}^{\infty} (x - d_{[i]}) \cdot (y + x - d_{[j]}) f_X(x) \cdot f_Y(y) dy dx$$

Rewriting $E[L_{[i]}L_{[j]}]$, we have

$$E[L_{[i]}L_{[j]}] = \int_{x=-\infty}^{\infty} \int_{y=-\infty}^{\infty} (x - d_{[i]}) \cdot [(y - \mu_Y) + (x - d_{[j]} + \mu_Y)]$$

$$\times f_X(x) \cdot f_Y(y) dy dx$$

$$= \int_{x=-\infty}^{\infty} (x - d_{[i]}) \cdot (x - d_{[j]} + \mu_Y) f_X(x) \int_{y=-\infty}^{\infty} f_Y(y) dy dx$$

$$+ \int_{x=-\infty}^{\infty} (x - d_{[i]}) \cdot f_X(x) \int_{y=-\infty}^{\infty} (y - \mu_Y) f_Y(y) dy dx$$

$$= \int_{x=-\infty}^{\infty} (x - d_{[i]}) \cdot (x - d_{[j]} + \mu_Y) f_X(x) dx$$

$$= \int_{x=-\infty}^{\infty} [x^2 + (\mu_Y - d_{[i]} - d_{[j]})x + d_{[i]}d_{[j]} - d_{[i]}\mu_Y] f_X(x) dx$$

Since $\int_{x=-\infty}^{\infty} x^2 f_X(x) dx = \sigma_X^2 + \mu_X^2$, $\int_{x=-\infty}^{\infty} x \cdot f_X(x) dx = \mu_X$, and $\int_{x=-\infty}^{\infty} f_X(x) dx = 1$, we have

$$E[L_{[i]}L_{[j]}] = \sigma_X^2 + \mu_X^2 + (\mu_Y - d_{[i]} - d_{[j]})\mu_X + d_{[i]}d_{[j]} - d_{[i]}\mu_Y$$

By the definition of coefficient of correlation,

$$\sigma_{L[i]}\sigma_{L[j]}\rho_{[ij]} = E[L_{[i]}L_{[j]}] - \mu_{L[i]}\mu_{L[j]}$$

By substituting the expression for $E[L_{[i]}L_{[j]}]$ from above, we have

$$\sigma_{L[i]}\sigma_{L[j]}\rho_{[ij]} = \sigma_X^2 + \mu_X^2 + (\mu_Y - d_{[i]} - d_{[j]})\mu_X + d_{[i]}d_{[j]} - d_{[i]}\mu_Y$$

$$- (\mu_X - d_{[i]}) \cdot (\mu_X + \mu_Y - d_{[i]})$$

$$= \sigma_X^2 = \sigma_{L[i]}^2$$

Therefore,

$$a_{[ij]}^2 = \sigma_{L[i]}^2 + \sigma_{L[j]}^2 - 2\sigma_{L[i]}\sigma_{L[j]}\rho_{[ij]} = \sigma_{L[j]}^2 - \sigma_{L[i]}^2$$

Using Clark's equations, the first moment, $E[\max(L_{[1]}, L_{[2]})]$, is given by

$$\nu_{1[12]} = \mu_{[12]} = \mu_{L[1]}\Phi(\alpha_{[12]}) + \mu_{L[2]}\Phi(-\alpha_{[12]}) + a_{[12]}\varphi(\alpha_{[12]})$$

and the second moment is given by

$$v_{2[12]} = (\mu_{L[1]}^2 + \sigma_{L[1]}^2)\Phi(\alpha_{[12]}) + (\mu_{L[2]}^2 + \sigma_{L[2]}^2)\Phi(-\alpha_{[12]})$$
$$+ (\mu_{L[1]} + \mu_{L[2]})a_{[12]}\varphi(\alpha_{[12]})$$

Since

$$\text{var}[\max(L_{[1]}, L_{[2]})] = v_{2[12]} - (v_{1[12]})^2$$

we have

$$\sigma_{[12]}^2 = (\mu_{L[1]}^2 + \sigma_{L[1]}^2)\Phi(\alpha_{[12]}) + (\mu_{L[2]}^2 + \sigma_{L[2]}^2)\Phi(-\alpha_{[12]})$$
$$+ (\mu_{L[1]} + \mu_{L[2]})a_{[12]}\varphi(\alpha_{[12]})$$
$$- (\mu_{L[1]}\Phi(\alpha_{[12]}) + \mu_{L[2]}\Phi(-\alpha_{[12]}) + a_{[12]}\varphi(\alpha_{[12]}))^2$$
$$\rho_{[123]} = r[L_{[3]}, \max(L_{[1]}, L_{[2]})] = [\sigma_{[1]}\rho_{[13]}\Phi(\alpha_{[12]}) + \sigma_{[2]}\rho_{[23]}\Phi(-\alpha_{[12]})]/\sigma_{[12]}$$

Next, consider $\max(L_{[1]}, L_{[2]}, L_{[3]})$. Thus

$$\max(L_{[1]}, L_{[2]}, L_{[3]}) = \max(\max(L_{[1]}, L_{[2]}), L_{[3]})$$
$$\rho_{[123]} = r[L_{[3]}, \max(L_{[1]}, L_{[2]})]$$
$$a_{[123]}^2 = \sigma_{[12]}^2 + \sigma_{L[3]}^2 - 2\sigma_{[12]}\sigma_{L[3]}\rho_{[123]}$$
$$\alpha_{[123]} = \frac{\mu_{[12]} - \mu_{L[3]}}{a_{[123]}}$$

The first moment of $E[\max(L_{[1]}, L_{[2]}, L_{[3]})]$ is given by

$$v_{1[123]} = \mu_{[123]} = \mu_{[12]}\Phi(\alpha_{[123]}) + \mu_{L[3]}\Phi(-\alpha_{[123]}) + a_{[123]}\varphi(\alpha_{[123]})$$

The second moment is given by

$$v_{2[123]} = (\mu_{[12]}^2 + \sigma_{[12]}^2)\Phi(\alpha_{[123]}) + (\mu_{L[3]}^2 + \sigma_{L[3]}^2)\Phi(-\alpha_{[123]})$$
$$+ (\mu_{[12]} + \mu_{L[3]})a_{[123]}\varphi(\alpha_{[123]})$$

Since $\sigma_{[123]}^2 = \text{var}[\max(L_{[1]}, L_{[2]}, L_{[3]})] = v_{2[123]} - (v_{1[123]})^2$, we have

$$\sigma_{[123]}^2 = (\mu_{[12]}^2 + \sigma_{[12]}^2)\Phi(\alpha_{[123]}) + (\mu_{L[3]}^2 + \sigma_{L[3]}^2)\Phi(-\alpha_{[123]})$$
$$+ (\mu_{[12]} + \mu_{L[3]})a_{[123]}\varphi(\alpha_{[123]})$$
$$- (\mu_{[12]}\Phi(\alpha_{[123]}) + \mu_{L[3]}\Phi(-\alpha_{[123]}) + a_{[123]}\varphi(\alpha_{[123]}))^2$$
$$\rho_{[1234]} = r[L_{[4]}, \max(L_{[1]}, L_{[2]}, L_{[3]})]$$
$$= [\sigma_{[12]}\rho_{[124]}\Phi(\alpha_{[123]}) + \sigma_{[3]}\rho_{[34]}\Phi(-\alpha_{[123]})]/\sigma_{[123]}$$

Proceeding in a similar way and applying Clark's equations recursively for $n-1$ steps, the expectation and variance of maximum lateness can be determined.

An evaluation of various sequences for this case, using the data given in Table 4.1, is as follows:

Sequence*:   4–2–3–1 with $E = 25.006$ and $V = 54.866$
Sequences**: 4–3–2–1 with $E = 26.545$ and $V = 45.113$
               3–4–2–1 with $E = 27.667$ and $V = 36.211$

Note that the two EV-efficient sequences have significantly smaller variances, with only slight increments in the expected value over the optimal value.

4.4 Concluding Remarks

In this chapter we have presented closed-form expressions (wherever possible) to compute the expectation and variance of various performance measures for sequencing a given set of jobs on a single machine. The completion-time-based measures that we considered are the total completion time, the total weighted completion time, and the total discounted weighted completion time. We developed generic expressions for each of these measures, and two separate cases were analyzed (involving exponentially and normally distributed job processing times) for the discounted weighted completion time measure.

The due-date-based measures considered include total tardiness, total weighted tardiness, total number of tardy jobs, total weighted number of tardy jobs, mean lateness, and maximum lateness. Generic expressions were developed for the mean lateness measure, whereas, for all the other measures, algorithms based on Clark's equations were devised to approximately evaluate the expectation and variance for normally distributed job processing times. In addition, for the maximum lateness measure, approximate evaluations were obtained under the assumption that the lateness random variables are mutually related by a multivariate normal distribution. We also have numerically illustrated the significance and applicability of the expressions and methodologies developed through an example problem.

5 Flow-Shop Models

5.1 Introduction

In Chapter 4, different single-machine models were dealt with in detail, and now we extend our expectation-variance (EV) analysis to multimachine environments. In this chapter we focus on a flow shop with unlimited intermediate storage (buffer capacity) where n jobs are waiting to be processed. The flow shop consists of m machines. Our analysis for this environment is identical to that presented by Wilhelm and Ahmadi-Marandi (1982), who studied an assembly system. A flow shop is, in fact, a simpler version of an assembly line, and hence their methodology can be adapted to our case. The EV analysis of a flow shop would act as a stepping stone for undertaking the EV analysis for job-shop and parallel-machine environments. Numerical illustrations indicating the significance and applicability of our work are also presented through example problems. The objective function that we consider is to minimize the makespan. Makespan is an important objective in a flow shop because a lower makespan value implies higher utilization of the machines. Also, we consider permutation flow shops, in which the job sequence does not change across the machines, and the jobs on each machine are also processed according to the first come, first served principle.

We adapt the notation that we have used so far for the flow-shop environment as follows:

m = number of machines

n = number of jobs

$P_{[i,j]}$ = processing time for the job in the jth position of the given permutation schedule on machine i (random variable)

$\mu_{[i,j]}$ = mean or expected value of processing time for the job in the jth position of the given permutation schedule on machine i

$\sigma^2_{[i,j]}$ = variance of the processing time for the job in the jth position of the given permutation schedule on machine i

$C_{[i,j]}$ = completion time for the job in the jth position of the given
        permutation schedule on machine i (random variable)

$S_{[i,j]}$ = start time for the job in the jth position of the given permutation
        schedule on machine i (random variable)

5.2 Permutation Flow Shops with Unlimited Intermediate Storage

Given a permutation schedule for an m-machine flow shop, the completion time for the first job in the schedule on all the machines can be computed easily and is given by

$$C_{[i,1]} = \sum_{k=1}^{i} P_{[k,1]}, \quad i = 1, \ldots, m$$

Similarly, the completion time for all the jobs on the first machine is

$$C_{[1,j]} = \sum_{l=1}^{j} P_{[1,l]}, \quad j = 1, \ldots, n$$

The completion times for all the other jobs on the machines then can be found recursively using the following expression:

$$C_{[i,j]} = \max\left(C_{[i-1,j]}, C_{[i,j-1]}\right) + P_{[i,j]}, \quad i = 2, \ldots, m; \ j = 2, \ldots, n$$

where $\max(C_{[i-1,j]}, C_{[i,j-1]})$ is the possible starting time of job j on machine i, and we denote it as $S_{i,j}$. The starting time for a job j on a machine i is determined by the completion time for job j on the previous machine $i-1$ and the completion time for the previous job $j-1$ in the sequence on machine i. We have

$$C_{[i,j]} = S_{[i,j]} + P_{[i,j]}, \quad i = 2, \ldots, m; \ j = 2, \ldots, n$$

The makespan of the given permutation schedule then would be given by the completion time of the last job on the last machine m in the given schedule. That is,

$$C_{\max} = C_{[m,n]} = S_{[m,n]} + P_{[m,n]} = \max\left(C_{[m-1,n]}, C_{[m,n-1]}\right) + P_{[m,n]}$$

5.2.1 Expectation and Variance of Makespan

Since there are no standard evaluations available for the maximum operator, it becomes necessary to use Clark's equations for approximate evaluation of the makespan under the assumption that all the start and finish times are mutually related by a multivariate normal distribution. $C_{\max}$ can be found by iteratively

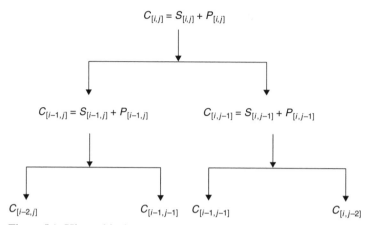

Figure 5.1. Hierarchical structure showing the association between completion times.

computing the earlier completion times and their correlations. Wilhelm and Ahmadi-Marandi (1982) have presented a seven-step iterative procedure to determine the correlations necessary to evaluate the maximum function for an assembly system. They incorporated stochastic part arrival times in their assembly-system model, whereas the processing times were assumed to be deterministic. A flow shop is a special case of an assembly system, and a similar approach can be adapted to determine the correlations and incorporate stochastic processing times. We hereby present a recursive procedure to determine the correlations that are then used to evaluate the mean and variance of the completion times.

The recursive procedure can be better understood from the following hierarchical structure of the completion times (Figure 5.1):

$$C_{[i,j]} = S_{[i,j]} + P_{[i,j]}$$

As expressed earlier, the completion time of the jth job on machine i is given by

$$C_{[i,j]} = S_{[i,j]} + P_{[ij]} = \max\left(C_{[i-1,j]}, C_{[i,j-1]}\right) + P_{[ij]}$$

The mean and variance of the completion times, $C_{[i-1,j]}$ and $C_{[i,j-1]}$, are known from the earlier iterations. The complexity lies in determining the crucial correlation factor between $C_{[i-1,j]}$ and $C_{[i,j-1]}$ denoted by $r(C_{[i-1,j]}, C_{[i,j-1]})$, which is required for use in Clark's equations. The starting times for operations $[i-1,j]$ and $[i,j-1]$, namely, $S_{[i-1,j]}$ and $S_{i,j-1}$, in turn, depend on the previous completion times, as shown in Figure 5.1.

At the beginning of every iteration, the following parameters are known:

$$r(C_{[i-1,j-1]}, C_{[i-1,j-1]}) = 1 \quad \text{(by definition)}$$

$$r(C_{[i-1,j-1]}, C_{[i,j-2]}) \qquad \text{(known from the previous iterations)}$$

$r(C_{[i-2,j]}, C_{[i-1,j-1]})$ (known from the previous iterations)

$r(C_{[i-2,j]}, C_{[i,j-2]}) = 0$ (as an assumption)

Although it is possible to determine $r(C_{[i-2,j]}, C_{[i,j-2]})$ by proceeding further down the hierarchy, it would only make the computations cumbersome and time consuming for larger problems. It is also reasonable to surmise that the correlations get weaker as we go further down the hierarchy because the machines and jobs get further apart from each other. Wilhelm (1986) has, in fact, analyzed the importance of correlations between these random variables in a flow shop and has identified a number of inherent relationships among them to compute better estimates of the transient performance of a flow shop with finite and infinite buffer capacities. However, we assume that $r(C_{[i-2,j]}, C_{[i,j-2]}) = 0$ and provide a four-step procedure to compute the mean and expectation for the completion times when the processing times are probabilistic.

Step 1. Determine the correlation $\rho_1 = r(C_{[i-2,j]}, C_{[i,j-1]})$.

$$\rho_1 = r(C_{[i-2,j]}, C_{[i,j-1]})$$

$$= \frac{\text{cov}[C_{[i-2,j]}, (S_{[i,j-1]} + P_{[i,j-1]})]}{\sqrt{\text{var}[C_{[i-2,j]}] \cdot \text{var}[C_{[i,j-1]}]}}$$

$$= \frac{\text{cov}[C_{[i-2,j]}, S_{[i,j-1]}]}{\sqrt{\text{var}[C_{[i-2,j]}] \cdot \text{var}[S_{[i,j-1]}]}} \cdot \frac{\sqrt{\text{var}[S_{[i,j-1]}]}}{\sqrt{\text{var}[C_{[i,j-1]}]}},$$

$$= r(C_{[i-2,j]}, \max(C_{[i-1,j-1]}, C_{[i,j-2]})) \frac{\sqrt{\text{var}[S_{[i,j-1]}]}}{\sqrt{\text{var}[C_{[i,j-1]}]}}$$

Hence the pairwise correlations between the completion times can be used in Clark's equations to calculate $r(C_{[i-2,j]}, \max(C_{[i-1,j-1]}, C_{[i,j-2]}))$ and then to determine $r(C_{[i-2,j]}, C_{[i,j-1]})$.

Step 2. Determine the correlation $\rho_2 = r(C_{[i-1,j-1]}, C_{[i,j-1]})$.

$$\rho_1 = r(C_{[i-1,j-1]}, C_{[i,j-1]})$$

$$= \frac{\text{cov}[C_{[i-1,j-1]}, (S_{[i,j-1]} + P_{[i,j-1]})]}{\sqrt{\text{var}[C_{[i-1,j-1]}] \cdot \text{var}[C_{[i,j-1]}]}}$$

$$= \frac{\text{cov}[C_{[i-1,j-1]}, S_{[i,j-1]}]}{\sqrt{\text{var}[C_{[i-1,j-1]}] \cdot \text{var}[S_{[i,j-1]}]}} \cdot \frac{\sqrt{\text{var}[S_{[i,j-1]}]}}{\sqrt{\text{var}[C_{[i,j-1]}]}},$$

$$= r(C_{[i-1,j-1]}, \max(C_{[i-1,j-1]}, C_{[i,j-2]})) \frac{\sqrt{\text{var}[S_{[i,j-1]}]}}{\sqrt{\text{var}[C_{[i,j-1]}]}}$$

Step 3. Determine the correlation $\rho = r(C_{[i-1,j]}, C_{[i,j-1]})$.

$$\rho = r(C_{[i-1,j]}, C_{[i,j-1]})$$

$$= \frac{\text{cov}[(S_{[i-1,j]} + P_{[i-1,j]})C_{[i,j-1]}]}{\sqrt{\text{var}[C_{[i-1,j]}] \cdot \text{var}[C_{[i,j-1]}]}}$$

$$= \frac{\text{cov}[S_{[i-1,j]}, C_{[i,j-1]}]}{\sqrt{\text{var}[S_{[i-1,j]}] \cdot \text{var}[C_{[i,j-1]}]}} \cdot \frac{\sqrt{\text{var}[S_{[i-1,j]}]}}{\sqrt{\text{var}[C_{[i-1,j]}]}}$$

$$= r(\max(C_{[i-2,j]}, C_{[i-1,j-1]}), C_{[i,j-1]}) \cdot \frac{\sqrt{\text{var}[S_{[i-1,j]}]}}{\sqrt{\text{var}[C_{[i-1,j]}]}}$$

The pairwise correlations obtained during steps 1 and 2 can be used to determine ρ.

Step 4. Determine the mean and variance of $C_{[i,j]}$. Once $\rho = r(C_{[i-1,j]}, C_{[i,j-1]})$ is known, the mean and variance for the starting time of operation $[i,j]$, $S_{[i,j]}$, then can be found using the first two moments from Clark's equations. The correlation ρ will be used in the successive iterations for evaluating other completion times.

The mean and variance for the completion time $C_{[i,j]}$ are then given by

$$E[C_{[i,j]}] = E[S_{[i,j]}] + E[P_{[i,j]}] = E[S_{[i,j]}] + \mu_{[i,j]}$$
$$\text{var}[C_{[i,j]}] = \text{var}[S_{[i,j]}] + \text{var}[P_{[i,j]}] = \text{var}[S_{[i,j]}] + \sigma^2_{[i,j]}$$

The iterative procedure begins for $C_{[2,2]}$, and the correlations $r(C_{[1,1]}, C_{[2,0]})$ and $r(C_{[0,2]}, C_{[1,1]})$ are assumed to be zero. The terms $C_{[0,j]}$ and $C_{[i,0]}$ can be construed as job arrival times and machine ready times, respectively, which are deterministic, and hence any correlation involving them is set to zero.

Consider the following instance involving four jobs and four machines. The mean times for processing the jobs on the machines are given in Table 5.1. The corresponding variances of the processing times are given in Table 5.2.

Table 5.1. *Mean Processing Times – Data Set I*

Job Index	1	2	3	4
Machine 1	10	2	5	48
Machine 2	8	3	9	5
Machine 3	2	11	7	2
Machine 4	2	2	8	4

Table 5.2. *Variance of Processing Times – Data Set I*

Job Index	1	2	3	4
Machine 1	3	1	2	2
Machine 2	1	2	2	1
Machine 3	2	1	2	1
Machine 4	2	2	5	3

Table 5.3. *Mean Processing Times – Data Set II*

Job Index	1	2	3	4	5	6	7	8	9	10
Machine 1	90	18	45	36	18	18	36	45	18	90
Machine 2	72	27	81	45	27	27	45	81	27	72
Machine 3	9	99	63	18	9	9	18	63	99	9
Machine 4	18	18	72	36	9	9	36	72	18	18
Machine 5	90	72	9	18	9	9	18	9	72	90
Machine 6	108	18	45	36	18	18	36	45	18	90
Machine 7	63	36	81	99	36	27	45	36	27	72
Machine 8	9	99	0	18	72	36	18	63	99	9
Machine 9	27	36	72	36	99	0	36	72	27	18
Machine 10	90	72	99	18	9	36	18	9	72	90

Table 5.4. *Variance of Processing Times – Data Set II*

Job Index	1	2	3	4	5	6	7	8	9	10
Machine 1	9	3	6	6	3	3	6	6	3	9
Machine 2	3	6	6	3	0	0	3	6	6	3
Machine 3	6	3	6	3	0	0	3	6	3	6
Machine 4	6	6	15	9	6	6	9	15	6	6
Machine 5	9	3	6	6	3	3	6	6	3	9
Machine 6	3	3	6	3	0	0	3	6	3	6
Machine 7	0	6	6	3	0	0	3	6	6	3
Machine 8	9	3	6	6	3	3	6	6	3	9
Machine 9	6	6	15	9	6	6	9	15	6	6
Machine 10	3	3	6	6	3	3	6	6	3	12

The results of the application of this procedure are shown below:

Sequence*: $2–4–3–1$ with $E = 38.485$ and $V = 12.145$

Sequences**: $2–3–4–1$ with $E = 38.592$ and $V = 11.994$

               $2–3–1–4$ with $E = 39.224$ and $V = 10.380$

               $3–4–2–1$ with $E = 40.094$ and $V = 10.083$

               $4–3–1–2$ with $E = 44.438$ and $V = 9.974$

               $3–4–1–2$ with $E = 44.620$ and $V = 9.728$

Consider another instance involving 10 jobs and 10 machines. The mean processing times are shown in Table 5.3, whereas the corresponding variances are shown in Table 5.4. The expected value and variance of the makespan for

Table 5.5. *Expectation and Variance of the Makespan for Various Sequences – Data Set II*

Sequence Number	Sequence										E	V
1	10	9	8	2	3	4	6	5	1	7	1188.69	97.02
2	1	2	10	9	4	8	6	7	5	3	1251.03	107.50
3	4	3	1	6	2	10	9	7	5	8	1138.40	95.38
4	1	2	3	4	5	6	7	8	9	10	1180.26	79.80
5	10	9	8	7	6	5	4	3	2	1	1224.01	110.95
6	2	1	4	6	8	10	5	7	9	3	1181.57	81.37
7	2	5	1	7	4	6	9	3	8	10	1108.71	77.85
8	8	1	3	5	4	7	10	6	9	2	1242.01	65.95
9	6	1	9	7	3	8	4	2	10	5	1152.19	85.69
10	5	9	3	10	1	6	8	2	7	4	1046.46	86.71

several sequences are shown in Table 5.5. Note that among these 10 examples, sequence 10 gives the smallest expectation, whereas the smallest variance is achieved for sequence 8.

5.3 Concluding Remarks

In this chapter we have analyzed the expectation and variance of the makespan of a schedule that is implemented on a flow shop with unlimited buffer capacity. We also have numerically illustrated the significance and applicability of the expressions and methodologies developed for determining the expectation and variance of the makespan in the flow shop using both a small-size and a large-size problem.

6 Job-Shop Models

6.1 Introduction

In this chapter we consider the job-shop machine configuration. The classic job-shop problem differs from the flow-shop problem by the fact that the jobs do not follow the same route; that is, the flow of work is not unidirectional. The system consists of m machines and n jobs, and each job follows its own sequence of operations over the machines. It is not necessary for the jobs to visit all the machines. If a job has to visit a machine more than once before its completion, then it is said to *recirculate*. Recirculation is quite common in semiconductor manufacturing environments and is a complex phenomenon to analyze. In our analysis, we focus on the classic job-shop problem with makespan objective and no recirculation. However, our approach can be extended to the recirculation case as well, as we show later.

The notation that we use here is identical to that used in Chapter 5 except that $P_{[i,j]}, \mu_{[i,j]}, \sigma^2_{[i,j]}, C_{[i,j]}$, and $S_{[i,j]}$ now correspond to the job in the jth position on machine i of the given schedule. In addition, we require the following notation:

$$S^i = \text{processing sequence for jobs on machine } i \text{ in the given schedule}$$
$$\text{(job indices)}$$
$$N^i = n(S^i) = \text{number of jobs processed on machine } i$$
$$R^k = \text{processing sequence for job } k \text{ (machine indices)}$$
$$\text{Job}(i,j) = \text{function that returns the index of the job in the } j\text{th position in}$$
$$\text{the sequence of the jobs on machine } i$$
$$\text{Pos}(i,k) = \text{function that returns the position of job } k \text{ in the sequence of the}$$
$$\text{jobs on machine } i$$
$$\text{Pre}(i,k) = \text{function that returns the index of the machine preceding}$$
$$\text{machine } i \text{ in the processing sequence for job } k$$

Subscripts written without brackets refer to the particular index and are not indicative of the positions. For example, C_{ik} represents the completion time of job k on machine i and does not refer to the position of k in machine i's sequence.

6.2 Job Shops with Unlimited Intermediate Storage and No Recirculation

The makespan, defined as $C_{\max} = \max\left(C_{\max}^1, C_{\max}^2, \ldots, C_{\max}^m\right)$, is equivalent to the completion time of the last job to leave the system, where $C_{\max}^i$ is the completion time of the last job on machine i.

$$C_{\max}^i = C_{[i,N^i]} = S_{[i,N^i]} + P_{[i,N^i]}, \quad i = 1, 2, \ldots, m \quad (6.1)$$

However, the starting time for the last job on machine i depends on the completion time for the previous job on machine i and also on the completion time for that job on the preceding machine in its sequence.

In general, the completion time for the job in the jth position on machine i is given by

$$C_{[i,j]} = S_{[i,j]} + P_{[i,j]}, \quad i = 1, 2, \ldots, m \text{ and } j = 1, 2, \ldots, n \quad (6.2)$$

However, $S_{[i,j]}$ is further given by

$$S_{[i,j]} = \max(C_{[i,j-1]}, C_{[p,q]}) \quad (6.3)$$

where, $p = \text{Pre}(i, \text{Job}(i,j))$ and $q = \text{Pos}(p, \text{Job}(i,j))$, for $i = 1, 2, \ldots, m$ and $j = 1, 2, \ldots, n$. Equations (6.2) and (6.3) have to be used recursively to evaluate Equation (6.1).

The expectation and variance of the completion time for the job in the jth position on machine i then is given by

$$E[C_{[i,j]}] = E[S_{[i,j]}] + E[P_{[i,j]}] = E[\max(C_{[i,j-1]}, C_{[p,q]})] + \mu_{[i,j]} \quad (6.4)$$

$$\text{var}[C_{[i,j]}] = \text{var}[S_{[i,j]}] + \text{var}[P_{[i,j]}] = \text{var}[\max(C_{[i,j-1]}, C_{[p,q]})] + \sigma_{[i,j]}^2 \quad (6.5)$$

6.2.1 Expectation and Variance of Makespan

As was noted earlier for flow shops, the maximum operator has to be evaluated using Clark's equations for approximate evaluation of the makespan under the assumption that all the start and finish times are mutually related by a multivariate normal distribution. However, the computational procedure is much more tedious than it is for flow shops.

The correlation structure for the flow shop follows a generic pattern, and it is easier to understand and evaluate the pairwise correlations. Additionally, assumption with regard to the correlation $r(C_{[i-2,j]}, C_{[i,j-2]})$ (see Section 5.2.1) made the computations relatively simpler. On the other hand, the job-shop correlation structure is much more complicated and involves complete enumeration of all the correlations to compute the means and variances of the completion times. An illustrative example is provided below to show the steps involved in the computational procedure for this case.

6.2.1.1 Completion Times

Consider a three-machine, three-job job shop with the following job routings:

Job 1: Machines $a-b-c$
Job 2: Machines $c-b-a$
Job 3: Machines $b-a-c$

The sequence of computations is shown for an arbitrary schedule given below:

Machine a: Jobs $1-2-3$
Machine b: Jobs $3-1-2$
Machine c: Jobs $2-1-3$

The Gantt chart shown in Figure 6.1 is drawn for the given schedule. It is not possible to ascertain the completion times in this case because the processing times are probabilistic. Hence Figure 6.1 shows a schematic of the relative positions of the jobs.

For the given schedule, jobs 1, 2, and 3 can start their first operations on machines a, c, and b, respectively, at time $t = 0$, and they can be referred to as the *initial operations*. These operations are independent of each other, and hence there exists no correlation among them. However, for other subsequent operations, it is necessary to recursively compute the correlations until the *initial operations* are reached. This is illustrated below:

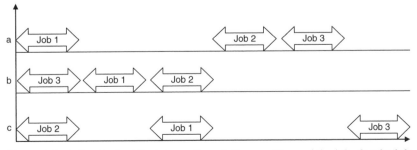

Figure 6.1. A Gantt chart that shows the relative positions of the jobs for the job-shop example.

Operation $(b, 1)$:

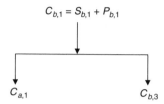

Since operations $(a, 1)$ and $(b, 3)$ are initial operations and independent, $r(C_{a,1}, C_{b,3}) = 0$.

Operation $(b, 2)$:

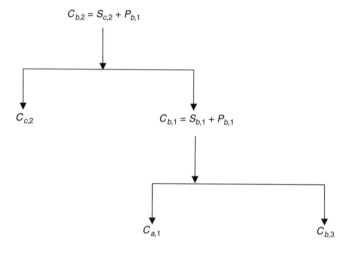

$r(C_{c,2}, C_{b,1}) = r(C_{c,2}, \max(C_{a,1}, C_{b,3}) + P_{b,1}) = 0$ because the three operations involved are all initial operations.

Operation $(c, 1)$:

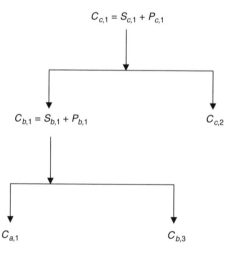

$r(C_{b,1}, C_{c,2}) = r(\max(C_{a,1}, C_{b,3}) + P_{b,1}, C_{c,2}) = 0$, as computed from the earlier step.

Operation $(a, 2)$:

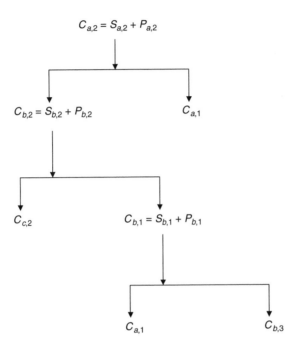

The steps involved in computing $r(C_{b,2}, C_{a,1})$ are as follows:

1. $r(C_{b,2}, C_{a,1}) = r(S_{b,2} + P_{b,2}, C_{a,1}) = r(\max(C_{c,2}, C_{b,1}), C_{a,1}) \cdot \sqrt{\dfrac{\mathrm{var}(S_{b,2})}{\mathrm{var}(C_{b,2})}}$

$$(6.6)$$

2. Consider the individual correlations $r(C_{c,2}, C_{a,1})$ and $r(C_{b,1}, C_{a,1})$. $r(C_{c,2}, C_{a,1}) = 0$ (since they are initial operations), and

$$r(C_{b,1}, C_{a,1}) = r(S_{b,1} + P_{b,1}, C_{a,1}) = r(\max(C_{a,1}, C_{b,3}), C_{a,1}) \cdot \sqrt{\dfrac{\mathrm{var}(S_{b,1})}{\mathrm{var}(C_{b,1})}},$$

   which can be found from Clark's equations using $r(C_{a,1}, C_{a,1}) = 1$ and $r(C_{a,1}, C_{b,3}) = 0$.
3. Using these correlations in the earlier steps, $r(C_{b,2}, C_{a,1})$ finally can be found.

Operation $(a, 3)$:

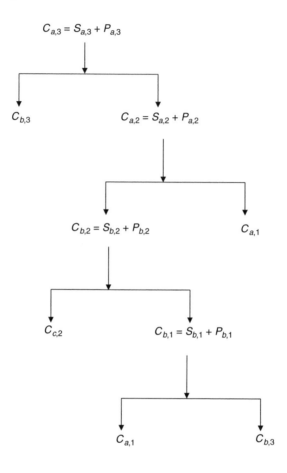

The required correlation $r(C_{b,3}, C_{a,2})$ can be found having known the pairwise correlations between $C_{b,3}, C_{b,2}$, and $C_{a,1}$. The correlations $r(C_{b,2}, C_{a,1})$ (from the previous iteration) and $r(C_{b,3}, C_{a,1})$ ($=0$, initial operations) are known, and the only unknown correlation is $r(C_{b,3}, C_{b,2})$. It can be found separately using a similar approach.

Operation $(c, 3)$: (see diagram on the next page)

The final correlation $r(C_{a,3}, C_{c,1})$ can be found by using the preceding structure and following the line of approach described earlier.

In general, the recursive procedure to compute mean and variance of the completion times of the various operations processed in accordance with a given schedule for a job shop can be summarized as follows:

Step 1. Identify the jobs that can be started at time $t = 0$, and designate them as initial operations. Compute the mean and variance of their completion times. Set the correlation coefficients between them to zero.

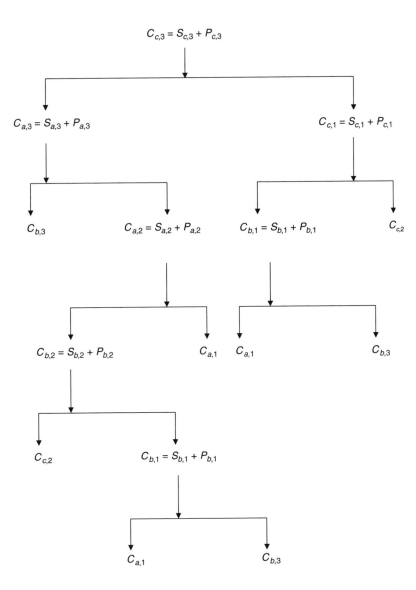

Step 2. Identify the next schedulable job k for the given schedule, trace the preceding operations that affect its starting time, and develop the hierarchical structure. The structure is expanded until the initial operations are reached.

Step 3. Find the correlation coefficients recursively, and determine the mean and variance for the operation start time.

Step 4. Compute the mean and variance of its completion time using the relation

$$C_{i,k} = S_{i,k} + P_{i,k}$$

Step 5. Repeat steps 2 through 4 until all the jobs are scheduled.

6.2.1.2 Determining the Makespan

After computing the means and variances of the completion times for all the operations, the makespan of the given schedule can be evaluated by using Clark's equations recursively in a manner similar to finding the maximum lateness for the single-machine model. Note that the makespan is

$$C_{max} = \max\left(C_{max}^1, C_{max}^2, \ldots, C_{max}^m\right) \quad \text{and}$$

$$C_{max}^i = C_{[i,N^i]} = S_{[i,N^i]} + P_{[i,N^i]}, \quad i = 1, 2, \ldots, m$$

The maximum function of m arguments can be broken recursively into $m - 1$ maximum functions as follows:

$$C_{max} = \max\left(C_{max}^1, C_{max}^2, \ldots, C_{max}^{m-1}, C_{max}^m\right)$$

$$= \max\left(\max\left(C_{max}^1, C_{max}^2, \ldots, C_{max}^{m-1}\right), C_{max}^m\right)$$

$$\vdots$$

$$= \max\left(\max\left(\ldots \max\left(\max\left(C_{max}^1, C_{max}^2\right), C_{max}^3\right), \ldots, C_{max}^{m-1}\right), C_{max}^m\right)$$

$$(6.7)$$

For machine i, the mean and variance of C_{max}^i is the mean and variance of the completion time of the last job in its sequence that has been determined earlier:

$$\mu_{C[i]} = \mathrm{E}\left[C_{[i,N^i]}\right] \quad \text{and} \quad \sigma^2_{C[i]} = \mathrm{var}\left[C_{[i,N^i]}\right]$$

Given that the processing times of the jobs are normally distributed and that the C_{max}^i values are mutually related by a multivariate normal distribution, C_{max} can be found as follows:

The correlation $\rho_{[ij]}$ is the correlation between the completion times of the last jobs on machine i and machine j, $r(C_{[i,N^i]}, C_{[j,N^i]})$. This can be computed, if the answers do not exist from earlier computations, using the correlation procedure given in the preceding discussion.

The expectation and variance of $\max\left(C_{max}^1, C_{max}^2\right)$ are computed using Clark's equations as given below. The first moment, $\mathrm{E}\left[\max\left(C_{max}^1, C_{max}^2\right)\right]$ is given by

$$\nu_{1[12]} = \mu_{[12]} = \mu_{C[1]}\Phi(\alpha_{[12]}) + \mu_{C[2]}\Phi(-\alpha_{[12]}) + a_{[12]}\varphi(\alpha_{[12]})$$

The second moment is given by

$$v_{2[12]} = \left(\mu^2_{C[1]} + \sigma^2_{C[1]}\right)\Phi(\alpha_{[12]}) + \left(\mu^2_{C[2]} + \sigma^2_{C[2]}\right)\Phi(-\alpha_{[12]})$$
$$+ \left(\mu_{C[1]} + \mu_{C[2]}\right) a_{[12]}\varphi(\alpha_{[12]})$$

$$\mathrm{var}\left[\max\left(C^1_{\max}, C^2_{\max}\right)\right] = v_{2[12]} - (v_{1[12]})^2 \quad \text{or}$$

$$\sigma^2_{[12]} = \left(\mu^2_{C[1]} + \sigma^2_{C[1]}\right)\Phi(\alpha_{[12]}) + \left(\mu^2_{C[2]} + \sigma^2_{C[2]}\right)\Phi(-\alpha_{[12]})$$
$$+ \left(\mu_{C[1]} + \mu_{C[2]}\right) a_{[12]}\varphi(\alpha_{[12]})$$
$$- \left(\mu_{C[1]}\Phi(\alpha_{[12]}) + \mu_{C[2]}\Phi(-\alpha_{[12]}) + a_{[12]}\varphi(\alpha_{[12]})\right)^2$$

$$\rho_{[123]} = r\left[C^3_{\max}, \max\left(C^1_{\max}, C^2_{\max}\right)\right]$$
$$= \left[\sigma_{C[1]}\rho_{[13]}\Phi(\alpha_{[12]}) + \sigma_{C[2]}\rho_{[23]}\Phi(-\alpha_{[12]})\right]/\sigma_{[12]}$$

Next, consider $\max\left(C^1_{\max}, C^2_{\max}, C^3_{\max}\right)$.

$$\max\left(C^1_{\max}, C^2_{\max}, C^3_{\max}\right) = \max\left(\max\left(C^1_{\max}, C^2_{\max}\right), C^3_{\max}\right)$$
$$\rho_{[123]} = r\left[C^3_{\max}, \max\left(C^1_{\max}, C^2_{\max}\right)\right]$$
$$a^2_{[123]} = \sigma^2_{[12]} + \sigma^2_{C[3]} - 2\sigma_{[12]}\sigma_{C[3]}\rho_{[123]}$$
$$\alpha_{[123]} = \frac{\mu_{[12]} - \mu_{C[3]}}{a_{[123]}}$$

The first moment $\mathrm{E}\left[\max\left(C^1_{\max}, C^2_{\max}, C^3_{\max}\right)\right]$, is given by

$$v_{1[123]} = \mu_{[123]} = \mu_{[12]}\Phi(\alpha_{[123]}) + \mu_{C[3]}\Phi(-\alpha_{[123]}) + a_{[123]}\varphi(\alpha_{[123]})$$

The second moment is given by

$$v_{2[23]} = \left(\mu^2_{[12]} + \sigma^2_{[12]}\right)\Phi(\alpha_{[123]}) + \left(\mu^2_{C[3]} + \sigma^2_{C[3]}\right)\Phi(-\alpha_{[13]})$$
$$+ \left(\mu_{[12]} + \mu_{C[3]}\right) a_{[123]}\varphi(\alpha_{[123]})$$

$$\mathrm{var}\left[\max\left(C^1_{\max}, C^2_{\max}, C^3_{\max}\right)\right] = v_{2[123]} - (v_{1[123]})^2 \quad \text{or}$$

$$\sigma^2_{[123]} = \left(\mu^2_{[12]} + \sigma^2_{[12]}\right)\Phi(\alpha_{[123]}) + \left(\mu^2_{C[3]} + \sigma^2_{C[3]}\right)\Phi(-\alpha_{[123]})$$
$$+ \left(\mu_{[12]} + \mu_{C[3]}\right) a_{[123]}\varphi(\alpha_{[123]})$$
$$- \left(\mu_{[12]}\Phi(\alpha_{[123]}) + \mu_{C[3]}\Phi(-\alpha_{[123]}) + a_{[123]}\varphi(\alpha_{[123]})\right)^2$$

$$\rho_{[1234]} = r\left[C^4_{\max}, \max\left(C^1_{\max}, C^2_{\max}, C^3_{\max}\right)\right]$$
$$= \left[\sigma_{[12]}\rho_{[124]}\Phi(\alpha_{[123]}) + \sigma_{C[3]}\rho_{[34]}\Phi(-\alpha_{[123]})\right]/\sigma_{[123]}$$

Thus, proceeding in a similar way and applying Clark's equations recursively for $m - 1$ steps, the expectation and variance of the makespan can be determined.

Example 6.1. Application of the Expectation and Variance Expressions for the Case of a Job Shop with Unlimited Buffer and without Circulation.

Consider the following instance with three jobs and three machines. The mean and variance of the processing times are given in Tables 6.1 and 6.2, respectively. The job routings for all the jobs are given in Table 6.3. The processing sequence for the three machines is given in Table 6.4. The means and variances of the completion times of the three jobs on the three machines for the given schedule are given in Tables 6.5 and 6.6.

The mean and variance of the makespan for the given schedule are 177.86 and 68.51, respectively. Similarly, other schedules can be evaluated as well.

Next, we consider another instance involving 15 jobs and 11 machines. The mean processing times and their variances are given in Tables 6.7 and 6.8.

Table 6.1. *Mean Processing Times*

Job Index	1	2	3
Machine 1	28	35	7
Machine 2	42	11	34
Machine 3	16	80	43

Table 6.2. *Variance of Processing Times*

Job Index	1	2	3
Machine 1	10	22	3
Machine 2	25	6	15
Machine 3	5	35	12

Table 6.3. *Job Routing Data*

Operation Sequence	Machine Index		
Job 1	1	2	3
Job 2	3	2	1
Job 3	2	1	3

Table 6.4. *Machine Processing Sequence*

Processing Sequence	Job Index		
Machine 1	1	2	3
Machine 2	3	1	2
Machine 3	2	1	3

Table 6.5. *Mean Completion Times*

Job Index	1	2	3
Machine 1	28.00	127.86	134.86
Machine 2	76.28	92.86	34.00
Machine 3	97.86	80.00	177.86

Table 6.6. *Variance of Completion Times*

Job Index	1	2	3
Machine 1	10.00	53.51	58.51
Machine 2	37.66	31.51	15.00
Machine 3	30.51	35.00	68.51

Table 6.7. *Mean Processing Times*

Job Index	1	2	3	4	5	6	7	8	9	10	11	12	13	14	15
Machine 1	72	23	55	91	43	49	75	72	39	5	24	68	80	95	47
Machine 2	26	85	50	68	70	69	90	9	80	88	49	18	60	23	28
Machine 3	38	67	61	72	34	9	91	91	63	68	18	77	40	92	24
Machine 4	19	43	93	96	12	27	17	31	52	50	91	60	56	6	77
Machine 5	30	39	7	99	30	59	8	38	35	53	33	18	51	10	48
Machine 6	98	13	80	68	7	72	37	98	25	24	19	15	89	23	29
Machine 7	74	19	57	60	74	45	71	49	35	20	38	90	91	46	77
Machine 8	8	9	46	43	84	99	50	90	47	57	99	10	72	72	8
Machine 9	43	74	72	12	40	61	65	62	22	53	9	60	86	41	55
Machine 10	75	26	42	11	69	63	98	72	74	58	35	72	79	34	49
Machine 11	43	73	72	11	40	60	65	62	21	52	9	60	85	40	55

Table 6.8. *Variances of Processing Times*

Job Index	1	2	3	4	5	6	7	8	9	10	11	12	13	14	15
Machine 1	7	2	6	9	4	5	8	7	4	1	2	7	8	10	5
Machine 2	3	9	5	7	7	7	9	1	8	9	5	2	6	2	3
Machine 3	4	7	6	7	3	1	9	9	6	7	2	8	4	9	2
Machine 4	2	4	9	10	1	3	2	3	5	5	9	6	6	1	8
Machine 5	3	4	1	10	3	6	1	4	4	5	3	2	5	1	5
Machine 6	10	1	8	7	1	7	4	10	3	2	2	2	9	2	3
Machine 7	7	2	6	6	7	5	7	5	4	2	4	9	9	5	8
Machine 8	I	1	5	4	8	10	5	9	5	6	10	1	7	7	1
Machine 9	4	7	7	1	4	6	7	6	2	5	1	6	9	4	6
Machine 10	8	3	4	1	7	6	10	7	7	6	4	7	8	3	5
Machine 11	4	7	7	1	4	6	7	6	2	5	1	6	9	4	6

Table 6.9. *Job Routing*

	Operation Sequence (Machine Index)								
Job 1	8	10	1	7	5	9	3	6	2	4
Job 2	7	11	4	1	2	5	6	10	3	8
Job 3	2	4	6	5	1	3	7	9	10	8
Job 4	2	8	5	7	6	1	11	4	10	3
Job 5	8	3	9	6	2	7	4	1	10	5
Job 6	11	1	5	6	10	2	8	7	4	3
Job 7	7	3	9	2	10	5	8	1	6	4
Job 8	9	8	6	4	3	5	10	2	1	7
Job 9	5	1	10	6	8	4	3	11	7	2
Job 10	10	1	4	11	2	7	3	6	5	8
Job 11	8	4	5	6	3	7	1	10	2	9
Job 12	1	4	3	8	9	6	10	2	7	5
Job 13	10	2	4	7	3	11	8	1	6	5
Job 14	5	3	6	7	11	8	4	2	1	10
Job 15	3	6	10	9	1	7	4	8	2	5

Table 6.10. *Machine Processing Sequence*

	Processing Sequence (Job Index)													
Machine 1	12	9	6	10	2	3	15	1	14	4	11	13	7	5	8
Machine 2	3	4	13	2	5	7	14	6	10	12	15	1	8	9	11
Machine 3	15	14	5	7	3	13	12	2	11	8	1	9	10	4	6
Machine 4	3	2	13	10	12	11	14	8	9	15	4	5	6	1	7
Machine 5	14	9	6	3	4	2	11	1	7	8	15	13	10	12	5
Machine 6	15	14	3	5	6	8	9	2	4	11	12	1	13	10	7
Machine 7	2	7	14	13	4	3	1	15	11	10	5	6	9	12	8
Machine 8	1	5	8	4	11	14	9	12	13	6	7	15	3	2	10
Machine 9	8	5	15	7	3	12	1	11	—	—	—	—	—	—	—
Machine 10	13	10	15	1	9	6	2	7	3	12	14	8	4	11	5
Machine 11	6	2	14	10	13	4	9	—	—	—	—	—	—	—	—

The job routings are given in Table 6.9, whereas Table 6.10 depicts the given processing sequence of the jobs on each machine. The mean and variance of the completion times for all the operations are given in Tables 6.11 and 6.12, respectively, for the given schedule. The mean and variance values of the makespan obtained after applying the preceding procedure are 1030.2 and 79.7, respectively.

6.3 Job Shops with Unlimited Intermediate Storage and with Recirculation

Job shops with recirculation can be modeled in a fashion identical to that for the case with no recirculation. A recirculating job can be treated anew independent

Table 6.11. *Mean of Completion Times*

Operation Index	1	2	3	4	5	6	7	8	9	10
Job 1	8.00	261.00	404.92	525.60	555.60	626.60	681.45	779.45	805.45	836.54
Job 2	19.00	133.00	186.05	209.05	294.05	369.50	438.67	464.67	534.18	768.02
Job 3	50.00	143.00	223.66	230.92	285.92	346.92	451.60	523.60	605.05	759.02
Job 4	118.00	225.00	330.50	394.60	506.67	664.18	675.18	777.55	796.53	884.55
Job 5	92.00	150.00	190.00	230.66	364.05	736.10	789.55	886.19	955.19	1021.20
Job 6	60.00	156.00	215.00	302.67	398.00	547.18	655.01	781.10	816.90	893.55
Job 7	90.00	241.00	311.28	454.05	563.05	571.93	705.01	843.19	930.22	947.22
Job 8	62.00	182.00	400.67	480.18	643.34	681.34	784.29	814.47	958.19	1007.20
Job 9	45.00	107.00	335.00	425.67	472.68	533.03	744.45	765.45	816.33	899.42
Job 10	137.00	161.00	292.17	344.17	635.18	662.10	812.45	893.22	973.22	1030.20
Job 11	324.00	443.18	476.18	525.68	552.34	640.60	688.19	831.53	948.42	957.42
Job 12	68.00	352.17	463.96	483.07	583.60	598.60	678.29	696.29	906.33	991.22
Job 13	79.00	178.00	242.17	333.17	386.96	471.96	555.22	768.19	869.22	920.22
Job 14	10.00	116.00	139.00	185.00	225.00	396.00	449.18	478.18	573.18	712.29
Job 15	24.00	53.00	186.00	245.93	332.92	602.60	679.60	713.02	741.05	789.05

Table 6.12. *Variance of Completion Times*

Operation Index	1	2	3	4	5	6	7	8	9	10
Job 1	1.00	27.00	35.71	42.30	45.30	52.30	53.66	63.66	66.66	52.79
Job 2	2.00	13.00	17.43	19.43	28.43	28.62	40.47	43.47	41.35	61.56
Job 3	5.00	14.00	18.71	17.71	23.71	29.71	35.30	42.30	53.47	60.56
Job 4	12.00	22.00	24.62	29.30	47.47	58.43	59.43	56.28	50.39	71.50
Job 5	9.00	14.00	18.00	19.71	35.43	53.24	57.28	80.08	87.08	78.95
Job 6	6.00	16.00	22.00	26.47	40.00	46.43	49.81	58.24	57.27	72.50
Job 7	9.00	23.00	25.30	44.43	49.47	42.37	54.81	76.08	70.02	72.02
Job 8	6.00	18.00	36.47	46.46	51.49	55.49	60.18	67.14	87.08	92.08
Job 9	5.00	11.00	34.00	39.47	44.45	43.94	59.66	61.66	58.74	55.05
Job 10	14.00	16.99	27.47	32.47	55.43	46.24	66.66	66.02	74.02	80.02
Job 11	32.00	42.46	45.46	49.46	42.49	54.30	60.08	54.39	60.05	61.05
Job 12	7.00	33.47	41.17	41.46	48.30	50.30	50.18	52.18	67.74	76.02
Job 13	8.00	18.00	22.47	31.47	33.17	42.17	46.86	68.08	64.02	69.02
Job 14	1.00	11.00	13.00	18.00	22.00	39.00	43.46	39.43	49.43	53.18
Job 15	2.00	5.00	19.00	19.65	28.71	50.30	58.30	55.56	57.94	62.94

and irrespective of the fact that it was processed on the same machine earlier. In the earlier job-shop example, if job 1 visits machine *b* after completing its operation on machine *c*, it is said to *recirculate*. However, we can assume it to be a new job, and the methodology explained earlier is still applicable because this operation is independent of its previous operation on the same machine. It depends only on the completion time of that operation's predecessor on the same machine $(b, 2)$ (in this case) as well as its completion time on the previous

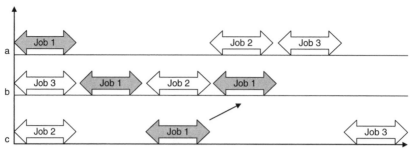

Figure 6.2. A Gantt chart for the job-shop example with recirculation.

machine $(c, 1)$. Hence the solution methodology for this case is identical to that for the no-recirculation case (see Figure 6.2).

6.4 Concluding Remarks

In this chapter we have analyzed the classic job-shop problem with unlimited buffer capacity and have detailed a generic recursive procedure for determining the correlation coefficients involved in computing the means and variances of operation completion times. This procedure is applicable for both the job recirculation and no-recirculation cases. The makespan then can be found by using Clark's equations recursively, which is also outlined. We have also numerically illustrated the significance and applicability of the expressions and methodologies developed for determining the expectation and variance of the makespan for the job-shop environment considered.

7 Parallel-Machine Models

7.1 Introduction

In this type of machine configuration, there are m similar machines in parallel, and there are n jobs to be processed on these machines. When dealing with parallel machines, the makespan becomes an objective of considerable interest as opposed to the single-machine case, where the makespan is invariant of job sequences. Preemptions also play a very important role in this case as compared with that for a single machine. *Preemption* implies that it is not necessary to continuously process a job on a machine, once started, until completion. The processing of a job can be interrupted, and the remaining work can be completed on the same machine or on any of the other machines. For the parallel-machines configuration, we consider preemption and no-preemption cases separately.

7.2 Parallel Machines with No Preemptions

The two performance measures considered for the no-preemption case are makespan and total completion time. The notation used here is identical to that used in Chapter 6.

7.2.1 Makespan with No Preemptions

The makespan, defined as $C_{\max} = \max\left(C_{\max}^1, C_{\max}^2, \ldots, C_{\max}^m\right)$, is equivalent to the completion time of the last job to leave the system. Let $C_{\max}^i$ be the completion time of the last job on machine i. That is,

$$C_{\max}^i = \sum_{j \in S^i} P_j = \sum_{j=1}^{N^i} P_{[i,j]}$$

Then $C_{\max}^1, C_{\max}^2, \ldots, C_{\max}^m$ are linear functions of random variables of job processing times. $C_{\max}$ is the greatest of a finite set of random numbers. For normally distributed variables, Clark's equations can be applied recursively to determine the expectation and variance of $C_{\max}$. If the processing times of the n jobs are independent and normally distributed, then $C_{\max}^1, C_{\max}^2, \ldots, C_{\max}^m$ are also normally distributed by the reproductive property of normal random variables.

Note that

$$C_{\max} = \max\left(C_{\max}^1, C_{\max}^2, \ldots, C_{\max}^m\right)$$
$$= \max\left(\max\left(C_{\max}^1, C_{\max}^2\right), \ldots, C_{\max}^m\right)$$

The mean and variance of the makespan of machine i is given by

$$E[C_{\max}^i] = \mu_{M[i]} = \sum_{j=1}^{N^i} \mu_{[i,j]}$$

$$\mathrm{var}[C_{\max}^i] = \sigma_{M[i]}^2 = \sum_{j=1}^{N^i} \sigma_{[i,j]}^2$$

The makespan of machine i, $C_{\max}^i$, is normally distributed with mean $\mu_{M[i]}$ and $\sigma_{M[i]}^2$. Moreover, the random variables $C_{\max}^1, C_{\max}^2, \ldots, C_{\max}^m$ are all independent of each other because no preemptions allowed, and hence the correlation coefficient between any pair among them is zero. That is,

$$\rho_{[ik]} = r\left(C_{\max}^i, C_{\max}^k\right) = 0, \quad \forall i, k = 1, 2, \ldots, m \text{ and } i \neq k \qquad (7.1)$$

The expectation and variance of $\max\left(C_{\max}^1, C_{\max}^2\right)$ are computed using Clark's equations as follows:

The first moment, $E\left[\max\left(C_{\max}^1, C_{\max}^2\right)\right]$, is given by

$$\nu_{1[12]} = \mu_{[12]} = \mu_{M[1]}\Phi(\alpha_{[12]}) + \mu_{M[2]}\Phi(-\alpha_{[12]}) + a_{[12]}\varphi(\alpha_{[12]})$$

where $a_{[12]}^2 = \sigma_{M[1]}^2 + \sigma_{M[2]}^2 - 2\sigma_{M[1]}\sigma_{M[2]}\rho_{[12]} = \sigma_{M[1]}^2 + \sigma_{M[2]}^2$ since $\rho_{[12]} = 0$ and

$$\alpha_{[12]} = \frac{\mu_{M[1]} - \mu_{M[2]}}{a_{12}}$$

The second moment is given by

$$\nu_{2[12]} = \left(\mu_{M[1]}^2 + \sigma_{M[1]}^2\right)\Phi(\alpha_{[12]}) + \left(\mu_{M[2]}^2 + \sigma_{M[2]}^2\right)\Phi(-\alpha_{[12]})$$
$$+ \left(\mu_{M[1]} + \mu_{M[2]}\right)a_{[12]}\varphi(\alpha_{[12]})$$

$$\mathrm{var}\left[\max\left(C_{\max}^1, C_{\max}^2\right)\right] = \nu_{2[12]} - (\nu_{1[12]})^2 \quad \text{or}$$

$$\sigma^2_{[12]} = \left(\mu^2_{M[1]} + \sigma^2_{M[1]}\right)\Phi(\alpha_{[12]}) + \left(\mu^2_{M[2]} + \sigma^2_{M[2]}\right)\Phi(-\alpha_{[12]})$$

$$+ \left(\mu_{M[1]} + \mu_{M[2]}\right)a_{[12]}\varphi(\alpha_{[12]})$$

$$- \left(\mu_{M[1]}\Phi(\alpha_{[12]}) + \mu_{M[2]}\Phi(-\alpha_{[12]}) + a_{[12]}\varphi(\alpha_{[12]})\right)^2$$

$$\rho_{[123]} = r\left[C^3_{max}, \max\left(C^1_{max}, C^2_{max}\right)\right]$$

$$= \left[\sigma_{M[1]}\rho_{[13]}\Phi(\alpha_{[12]}) + \sigma_{M[2]}\rho_{[23]}\Phi(-\alpha_{[12]})\right]/\sigma_{M[12]}$$

$$= 0 \tag{7.2}$$

because $\rho_{[13]} = 0$ and $\rho_{[23]} = 0$.

Next, consider $\max\left(C^1_{max}, C^2_{max}, C^3_{max}\right)$. The expectation and variance of $\max\left(C^1_{max}, C^2_{max}, C^3_{max}\right)$ are as follows:

$$\max\left(C^1_{max}, C^2_{max}, \ldots, C^3_{max}\right) = \max\left(\max\left(C^1_{max}, C^2_{max}\right), C^3_{max}\right)$$

$$r\left[\max\left(C^1_{max}, C^2_{max}\right), C^3_{max}\right] = \rho_{[123]} = 0$$

$$a^2_{[123]} = \sigma^2_{[12]} + \sigma^2_{M[3]} - 2\sigma_{[12]}\sigma_{M[3]}\rho_{[123]} = \sigma^2_{[12]} + \sigma^2_{M[3]}$$

$$\alpha_{[123]} = \frac{\mu_{[12]} - \mu_{M[3]}}{a_{[123]}}$$

The first moment, $E\left[\max\left(C^1_{max}, C^2_{max}, C^3_{max}\right)\right]$, is given by

$$v_{1[123]} = \mu_{[123]} = \mu_{M[12]}\Phi(\alpha_{[123]}) + \mu_{M[3]}\Phi(-\alpha_{[123]}) + a_{[123]}\varphi(\alpha_{[123]})$$

The second moment is given by

$$v_{2[123]} = \left(\mu^2_{M[12]} + \sigma^2_{M[12]}\right)\Phi(\alpha_{[123]}) + \left(\mu^2_{M[3]} + \sigma^2_{M[3]}\right)\Phi(-\alpha_{[123]})$$

$$+ \left(\mu_{M[12]} + \mu_{M[3]}\right)a_{[123]}\varphi(\alpha_{[123]})$$

$$\text{var}\left[\max\left(C^1_{max}, C^2_{max}, C^3_{max}\right)\right] = v_{2[123]} - (v_{1[123]})^2 \quad \text{or}$$

$$\sigma^2_{[123]} = \left(\mu^2_{M[12]} + \sigma^2_{M[12]}\right)\Phi(\alpha_{[123]}) + \left(\mu^2_{M[3]} + \sigma^2_{M[3]}\right)\Phi(-\alpha_{[123]})$$

$$+ \left(\mu_{M[12]} + \mu_{M[3]}\right)a_{[123]}\varphi(\alpha_{[123]})$$

$$- \left(\mu_{M[12]}\Phi(\alpha_{[123]}) + \mu_{M[3]}\Phi(-\alpha_{[123]}) + a_{[123]}\varphi(\alpha_{[123]})\right)^2$$

$$\rho_{[1234]} = r\left[C^3_{max}, \max\left(C^1_{max}, C^2_{max}, C^3_{max}\right)\right]$$

$$= \left[\sigma_{M[12]}\rho_{[124]}\Phi(\alpha_{[123]}) + \sigma_{[3]}\rho_{[34]}\Phi(-\alpha_{[123]})\right]/\sigma_{[123]}$$

Using the correlation Equations (7.1) and (7.2), $\rho_{[124]} = 0$ and $\rho_{[34]} = 0$, and hence $\rho_{[1234]} = 0$. Thus, in a similar way, Clark's equations can be used recursively in $m - 1$ steps to determine the expectation and variance of C_{max}.

Table 7.1. *Data Set for the Case of Parallel Machines*

Job Index	1	2	3	4	5	6	7	8
Mean	2	4	6	8	7	10	12	3
Variance	2	3	2	2	3	2	4	3

Table 7.2. *Expectation and Variance of the Makespan*

Schedule Number	Sequence of Each Machine	E[makespan]	Var[makespan]
1	Machine 1: 1–2–3 Machine 2: 4–5 Machine 3: 6–7–8	25.00	8.94
2	Machine 1: 1–2 Machine 2: 3–4–5 Machine 3: 6–7–8	25.33	7.24
3	Machine 1: 1–2–7–8 Machine 2: 3–4–5 Machine 3: 6	22.74	6.48
4	Machine 1: 2–4 Machine 2: 1–3–7 Machine 3: 8–5–6	21.60	5.39
5	Machine 1: 4–5 Machine 2: 3–6–7 Machine 3: 1–8–2	28.00	8.00

Example 7.1. Application of the Expectation and Variance Expressions for the Case of Parallel Machines without Preemption and the Objective of Minimizing the Makespan.

Consider an instance with eight jobs and three machines. The mean and variance of the processing times are given in Table 7.1.

The expected values and variances of the makespan for the various schedules on these machines are shown in Table 7.2.

7.2.2 Total Completion Time with No Preemptions

Let C^i be the total completion time for all the jobs assigned to machine i. The C^i values are random variables, being linear functions of random variables (job processing times), and are independent of each other because no preemptions are allowed in the schedule. Thus

$$C^i = \sum_{j \in S^i} C_{[i,j]} = \sum_{j=1}^{N^i} \sum_{i=1}^{j} P_{[i,j]} \qquad (7.3)$$

Rearranging and summing similar terms within the summation, we have

$$C^i = P_{[i,1]} + (P_{[i,1]} + P_{[i,2]}) + \cdots + (P_{[i,1]} + P_{[i,2]} + \cdots + P_{[i,N^i-1]})$$
$$+ (P_{[i,1]} + P_{[i,2]}) + \cdots + P_{[i,N^i]})$$

$$C^i = \sum_{j=1}^{N^i}(N^i + 1 - j)P_{[i,j]}$$

$$E[C^i] = E\left[\sum_{j=1}^{N^i}(N^i + 1 - j)P_{[i,j]}\right] = \sum_{j=1}^{N^i}(N^i + 1 - j)\mu_{[i,j]}$$

$$\text{var}[C^i] = \text{var}\left[\sum_{j=1}^{N^i}(N^i + 1 - j)P_{[i,j]}\right] = \sum_{j=1}^{N^i}(N^i + 1 - j)^2\sigma_{[i,j]}$$

Hence

$$E\left[\sum_{i=1}^{m} C^i\right] = \sum_{i=1}^{m} E[C^i] = \sum_{i=1}^{m}\sum_{j=1}^{N^i}(N^i + 1 - j)\mu_{[i,j]}$$

and

$$\text{var}\left[\sum_{i=1}^{m} C^i\right] = \sum_{i=1}^{m} \text{var}[C^i] = \sum_{i=1}^{m}\sum_{j=1}^{N^i}(N^i + 1 - j)^2\sigma_{[i,j]}^2$$

Table 7.3. *Expectation and Variance of the Total Completion Time (TCT)*

Schedule Number	Sequence of Each Machine	E[TCT]	var[TCT]
1	Machine 1: 1–2–3 Machine 2: 4–5 Machine 3: 6–7–8	100	80
2	Machine 1: 1–2 Machine 2: 3–4–5 Machine 3: 6–7–8	106	77
3	Machine 1: 1–2–7–8 Machine 2: 3–4–5 Machine 3: 6	98	109
4*	**Machine 1: 2–4** **Machine 2: 1–3–7** **Machine 3: 8–5–6**	**79**	**85**
5**	Machine 1: 4–5 Machine 2: 3–6–7 Machine 3: 1–8–2	89	74

Example 7.2. Application of the Expectation and Variance Expressions for the Case of Parallel Machines without Preemption and the Objective of Minimizing Total Completion Time.

Using the data given in Table 7.2, the expected values and variances of total completion times for various schedules are shown in Table 7.3, and the schedule associated with the best expectation value and the expectation-variance-efficient (EV-efficient) schedule are also indicated.

7.3 Parallel Machines with Preemptions

A scheduling problem with preemptions is relatively more complex to model. Preemptions allow jobs to be interrupted during processing, and the work remaining can be scheduled for processing later on the same machine or on any of the other parallel machine. In our EV analysis, we are given the sequence of jobs on each of the machines, and with preemptions being allowed, a job could be processed on multiple machines. An illustrative Gantt chart for such a schedule is shown in Figure 7.1.

A schematic representation of the relative positions for different jobs is shown because it is not possible to ascertain the completion times owing to the probabilistic nature of the jobs. Jobs 1, 3, and 8 are preempted, whereas the other jobs are processed without interruption. For all the preempted jobs, the starting time of an operation depends on the completion time of its previous portion as well as the completion time of its predecessor on the same machine. For example, the completion time of the second portion of job 8 (i.e., 8–2) on machine c is given by $C_{c,8-2} = S_{c,8-2} + P_{c,8-2}$, where $S_{c,8-2}$ and $P_{c,8-2}$ represent the starting time and processing time of job 8–2 on machine c, respectively. However, the starting time $S_{c,8-2}$, in turn, depends on the completion time of job 1–3 on machine c as well as on the completion time of job 8–1 on machine b. The starting time $S_{c,8-2}$ then can be given as $S_{c,8-2} = \max(C_{b,8-1}, C_{c,1-3})$.

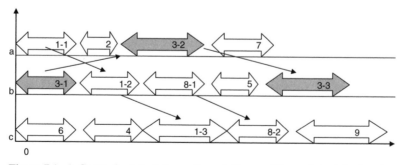

Figure 7.1. A Gantt chart that shows the relative positions of the jobs for the parallel machine with preemptions permitted.

Recursively, the completion time of job 1–3 on machine c is given by $C_{c,1-3} = S_{c,1-3} + P_{c,1-3}$, which, in turn, depends on the completion times of its earlier portion, job 1–2, and its predecessor, job 4, on machine c, and so on. Other preempted jobs also require similar recursive computations.

From the Gantt chart, we can infer that this case of parallel machines with preemptions is analogous to a job shop with recirculation and, possibly, that not all the jobs visit all the machines. A preempted job will have its own sequence for processing its preempted portions that is similar to the job processing routes in a job shop. Hence the same solution methodology can be applied here.

7.4 Concluding Remarks

In this chapter we have devised methodologies and determined closed-form expressions (wherever possible) to compute the expectation and variance of the makespan and total completion time for scheduling jobs on parallel machines with no preemptions. Generic expressions were developed for the case of total completion time, whereas the makespan analysis is applicable to normally distributed job processing times. The underlying solution methodology for the preemption case was found to be similar to that of a job shop with recirculation. Numerical illustrations through a few example problems also were provided to underline the significance and applicability of the expressions and methodologies developed.

8 The Case of General Processing Time Distribution

8.1 Introduction

In Chapters 4 through 7 we have developed expressions of expectation and variance for various performance measures of a given schedule of jobs. We have considered schedules for different machine configurations. For some of the performance measures of these schedules, necessary assumptions are made on the type of processing time distribution used to enable development of analytical expressions. Table 8.1 gives an overview of the machine configurations and performance measures that we have considered, and it also depicts the assumptions made on the processing time distributions used for each of these cases. This table also presents information on the level of accuracy of the resulting expressions for the expectation and variance of a performance measure and whether the analysis relies on Clark's method (Clark, 1961).

Clark's method for approximating the expectation and variance of the maximum of a set of random variables is based on the assumption of normal distributions for all random variables. In this chapter we relax this assumption and consider the case of general processing time distributions. Our analysis relies on the use of finite-mixture models.

8.1.1 Finite-Mixture Models

The use of a finite mixture of distributions provides a flexible methodology to represent a variety of random phenomena. In a mixture model, a random variable X is represented as a mixture of g components (random variables), X_i, $i = 1, 2, \ldots, g$. The random variable X takes the value of component X_i with probability p_i. Therefore, in the case of a continuous distribution, the probability density function (p.d.f) $f(x)$ of X can be written as

$$f(x) = \sum_{i=1}^{g} p_i f_i(x) \tag{8.1}$$

Table 8.1. *Assumptions on Processing Time Distributions and Accuracy of Approximation*

Machine Configuration	Performance Measure	Assumption on Processing Time Distribution	Use of Clark's Method	Level of Accuracy
Single machine	Total completion time	Default*	No	Accurate
Single machine	Total weighted completion time	Default	No	Accurate
Single machine	Total weighted discounted completion time	Moment-generating function is available	No	Accurate
Single machine	Total tardiness	Normal distribution	Yes	Accurate[†]
Single machine	Total weighed tardiness	Normal distribution	Yes	Accurate[†]
Single machine	Total number of tardy jobs	Normal distribution	No	Accurate[†]
Single machine	Total weighted number of tardy jobs	Normal distribution	No	Accurate[†]
Single machine	Mean lateness	Default	No	Accurate
Single machine	Maximum lateness	Normal distribution	Yes	Approximate
Flow shop	Makespan	Normal distribution	Yes	Approximate
Job shop	Makespan	Normal distribution	Yes	Approximate
Parallel machine	Makespan	Normal distribution	Yes	Approximate
Parallel machine	Total completion time	Default	No	Accurate

* No assumption required.
[†] Accuracy relies on that of the method of numerical integration used.

where $0 < p_i < 1$, and $\sum_{i=1}^{g} p_i = 1$. Note that $f_i(\cdot)$ is the p.d.f. of the ith component. The values of p_i, $i = 1,\ldots,g$, are called *mixing proportions* or *weights*.

Our motivation behind the use of a mixture of multiple random components is its ability to provide a number of controllable parameters, namely, those belonging to the weights and the shape of the component distributions, which allow the p.d.f. of X to take various shapes. We illustrate this point by a few examples (see Figure 8.1), as presented by Mclachlan and Peel (2001). Note that all these examples comprise normal components. Such mixtures are also called *normal mixtures*.

A mixture model provides a useful framework for representing a wide range of distributions having different shapes. As a result, it has found applications in astronomy, biology, genetics, medicine, psychiatry, economics, engineering, marketing, and other fields in the biological, physical, and social sciences (Mclachlan and Peel, 2001). Mixture models have been used to represent unknown distribution shapes and to explore the group structures of data. As noted by Mclachlan and Peel (2001), any continuous distribution can be approximated arbitrarily well by a finite mixture of normal densities with a

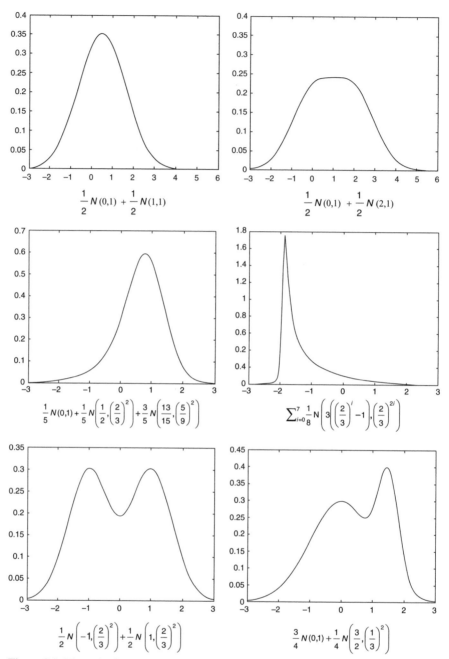

Figure 8.1. Normal mixture examples.

common variance (or covariance matrix in the multivariate case). This property enables an extension of our results presented in previous chapters. For a comprehensive coverage on the theory and applications of finite-mixture models, see Mclachlan and Peel (2001). Results on the asymptotic properties of maximum-likelihood estimators (MLEs) for mixture models have been provided by van de Geer (1996, 2003).

8.1.2 Maximum-Likelihood Fitting of Mixture Models

Given the number of components in a mixture model, there are several ways to fit it to a given random variable (or distribution). We can either use the method of moments, which can be computationally intensive because it requires derivatives of the functions involved, or we can use the expectation-maximization (EM) algorithm (Dempster et al., 1977). This algorithm consists of two steps, namely, the expectation (E) step and the maximization (M) step. Given the parameter values for the density functions used in the mixture and the corresponding weights and sample data, the E step estimates the sufficient statistics (in our case, the posterior probability that a data point belongs to a particular mixture component). The M step then determines parameter values by the maximum-likelihood method as if the estimated function values were the observed data. The EM algorithm converges to the MLEs if the model is identifiable. This method is very attractive computationally. The EM algorithm is easier to implement if all the components of the mixture have normal distributions. As in previous chapters, we also assume here that the processing times of different job operations are independent. Therefore, the approximation of processing time distributions can be performed on an operation-by-operation basis. In other words, given the (univariate) processing time distribution of an operation and the number of components in the mixture, our goal is to find the MLEs of the weights and the parameters for all the components.

To explicitly depict the distribution parameters, the p.d.f. of the mixture is rewritten as

$$f(x; \Psi) = \sum_{i=1}^{g} p_i f_i(x; \theta_i) \tag{8.2}$$

where θ_i is the vector of parameters for the ith component in the mixture, $i = 1, 2, \ldots, g$; and Ψ represents all the unknown parameters. That is, $\Psi = (p_1, \ldots, p_{g-1}, \theta_1, \ldots, \theta_g)$. Note that since $\sum_{i=1}^{g} p_i = 1$, only $(g-1)$ weights are considered as parameters. Suppose that m independent samples are generated from the processing time distribution and that these are denoted as x_j, $j = 1, 2, \ldots, m$. The log-likelihood value for Ψ can be calculated as

$$\log L(\Psi) = \sum_{j=1}^{m} \log f(x_j; \Psi)$$

$$= \sum_{j=1}^{m} \log \sum_{i=1}^{g} p_i f_i(x_j; \theta_i) \tag{8.3}$$

For computing the MLE of Ψ, we set $\partial \log L(\Psi)/\partial \Psi = 0$, which leads to

$$0 = \frac{\partial \log L(\Psi)}{\partial p_h} = \frac{\partial \left[\sum_{j=1}^{m} \log \sum_{i=1}^{g} p_i f_i(x_j; \theta_i) \right]}{\partial p_h} = \sum_{j=1}^{m} \frac{\partial \left[\log \sum_{i=1}^{g} p_i f_i(x_j; \theta_i) \right]}{\partial p_h}$$

$$= \sum_{j=1}^{m} \frac{\partial \left[\sum_{i=1}^{g} p_i f_i(x_j; \theta_i) \right] / \partial p_h}{\sum_{i=1}^{g} p_i f_i(x_j; \theta_i)}$$

$$= \sum_{j=1}^{m} \frac{\partial \left[\sum_{i=1}^{g-1} p_i f_i(x_j; \theta_i) + \left(1 - \sum_{i=1}^{g-1} p_i \right) f_g(x_j; \theta_g) \right] / \partial p_h}{\sum_{i=1}^{g} p_i f_i(x_j; \theta_i)}$$

$$= \sum_{j=1}^{m} \frac{f_h(x_j; \theta_h) - f_g(x_j; \theta_g)}{\sum_{i=1}^{g} p_i f_i(x_j; \theta_i)}, \qquad \forall h = 1, 2, \ldots, (g-1) \tag{8.4}$$

and

$$0 = \frac{\partial \log L(\Psi)}{\partial \theta_h} = \frac{\partial \left[\sum_{j=1}^{m} \log \sum_{i=1}^{g} p_i f_i(x_j; \theta_i) \right]}{\partial \theta_h} = \sum_{j=1}^{m} \frac{\partial \left[\log \sum_{i=1}^{g} p_i f_i(x_j; \theta_i) \right]}{\partial \theta_h}$$

$$= \sum_{j=1}^{m} \frac{p_h \frac{\partial f_h(x_j; \theta_h)}{\partial \theta_h}}{\sum_{i=1}^{g} p_i f_i(x_j; \theta_h)}, \qquad \forall h = 1, 2, \ldots, g \tag{8.5}$$

Let

$$\tau_h(x_j; \Psi) = \frac{p_h f_h(x_j; \theta_h)}{\sum_{i=1}^{g} p_i f_i(x_j; \theta_i)}, \qquad \forall h = 1, 2, \ldots, g \tag{8.6}$$

Note that $\tau_h(x_j; \Psi)$ is the posterior probability that x_j belongs to the hth component of the mixture. Consequently, $\forall h = 1, 2, \ldots, g$, we have

$$\frac{f_h(x_j; \theta_h)}{\sum_{i=1}^{g} p_i f_i(x_j; \theta_i)} = \frac{\tau_h(x_j; \Psi)}{p_h} \tag{8.7}$$

and

$$\frac{p_h}{\sum_{i=1}^{g} p_i f_i(x_j; \theta_i)} = \frac{\tau_h(x_j; \Psi)}{f_h(x_j; \theta_h)} \tag{8.8}$$

By Equations (8.4) and (8.7), we have

$$\sum_{j=1}^{m} \frac{\tau_h(x_j; \Psi)}{p_h} - \sum_{j=1}^{m} \frac{\tau_g(x_j; \Psi)}{p_g} = 0, \qquad \forall h = 1, 2, \ldots, (g-1) \tag{8.9}$$

If we let $K \equiv \sum_{j=1}^{m} \tau_g(x_j; \Psi)/p_g$, we have

$$\sum_{j=1}^{m} \tau_h(x_j; \Psi) = K \cdot p_h, \qquad \forall h = 1, 2, \ldots, g$$

Hence

$$K = K \sum_{h=1}^{g} p_h = \sum_{h=1}^{g} K \cdot p_h = \sum_{h=1}^{g} \sum_{j=1}^{m} \tau_h(x_j; \Psi) = \sum_{j=1}^{m} \sum_{h=1}^{g} \frac{p_h f_h(x_j; \theta_h)}{\sum_{i=1}^{g} p_i f_i(x_j; \theta_i)} = m$$

Consequently, we have

$$p_h = \frac{\sum_{j=1}^{m} \tau_h(x_j; \Psi)}{m}, \qquad \forall h = 1, 2, \ldots, g \tag{8.10}$$

On the other hand, by Equations (8.5) and (8.8), we have

$$\sum_{j=1}^{m} \tau_i(x_j; \Psi) \frac{\partial \log f_h(x_j; \theta_h)}{\partial \theta_h} = \sum_{j=1}^{m} \frac{\tau_h(x_j; \Psi)}{f_h(x_j; \theta_h)} \cdot \frac{\partial f_h(x_j; \theta_h)}{\partial \theta_h}$$

$$= \sum_{j=1}^{m} \frac{p_h}{\sum_{i=1}^{g} p_i f_i(x_j; \theta_i)} \cdot \frac{\partial f_h(x_j; \theta_h)}{\partial \theta_h} = 0, \quad \forall h = 1, 2, \ldots, g \tag{8.11}$$

As a part of the EM algorithm (Dempster et al., 1977), Equations (8.6), (8.10), and (8.11) can be solved iteratively to determine the MLE of a mixture model. To show this, we define a g-dimensional binary variable $\mathbf{z}_j$ for each sample x_j such that the ith component z_{ij} is 1 if and only if x_j belongs to the ith component of the mixture. Without loss of generality, we assume that the values of $\mathbf{z}_j$ follow a multinomial distribution $Z_j \approx \text{mult}_g(1, \mathbf{p})$. If the complete data (both x_j and $\mathbf{z}_j$) are available, the corresponding log likelihood for Ψ is given by

$$\log L_C(\Psi) = \sum_{j=1}^{m} \sum_{i=1}^{g} z_{ij}[\log p_i + \log f_i(x_j; \theta_i)] \tag{8.12}$$

Since the values of $\mathbf{z}_j$ are not observable, we treat them as missing data. As noted earlier, the EM algorithm consists of two iterative steps. In the E step, we overcome the difficulty of missing data ($\mathbf{z}_j$) by computing the sufficient statistics, namely the conditional expectation of z_{ij} based on the current estimation of distribution parameters $\Psi^{(k)}$. Since the conditional expectation

$$E_{\Psi^{(k)}}[Z_{ij}|x_j] = 1 \cdot P_{\Psi^{(k)}}\{Z_{ij} = 1|x_j\} = \frac{P_{\Psi^{(k)}}\{Z_{ij} = 1, X = x_j\}}{P_{\Psi^{(k)}}\{X = x_j\}}$$

$$= \frac{p_i^{(k)} f_i(x_j; \theta_i^{(k)})}{\sum_{h=1}^{g} p_h^{(k)} f_h(x_j; \theta_h^{(k)})} = \tau_i(x_j; \Psi^{(k)}) \tag{8.13}$$

we can calculate the conditional expectation of the log likelihood from Equation (8.12) as

$$E_{\Psi^{(k)}}[\log L_C(\Psi)|\mathbf{x}] = \sum_{j=1}^{m}\sum_{i=1}^{g} \tau_i(x_j;\Psi^{(k)})[\log p_i + \log f_i(x_j;\theta_i)] \qquad (8.14)$$

In the M step, we maximize the conditional expectation by varying the value of Ψ on the right-hand side of Equation (8.14). By the first-order optimality condition associated with p_h, we have

$$0 = \frac{\partial E_{\Psi^{(k)}}[\log L_C(\Psi)|\mathbf{x}]}{\partial p_h}$$

$$= \frac{\partial \left\{ \sum_{j=1}^{m}\sum_{i=1}^{g} \tau_i(x_j;\Psi^{(k)})[\log p_i + \log f_i(x_j;\theta_i)] \right\}}{\partial p_h}$$

$$= \sum_{j=1}^{m} \frac{\partial \left\{ \sum_{i=1}^{g} \tau_i(x_j;\Psi^{(k)})[\log p_i + \log f_i(x_j;\theta_i)] \right\}}{\partial p_h}$$

$$= \sum_{j=1}^{m} \left(\frac{\partial \left\{ \sum_{i=1}^{g-1} \tau_i(x_j;\Psi^{(k)})[\log p_i + \log f_i(x_j;\theta_i)] \right\}}{\partial p_h} \right.$$

$$\left. + \frac{\partial \left\{ \tau_g(x_j;\Psi^{(k)}) \left[\log\left(1 - \sum_{i=1}^{g-1} p_i\right) + \log f_g(x_j;\theta_g)\right] \right\}}{\partial p_h} \right)$$

$$= \sum_{j=1}^{m} \frac{\tau_h(x_j;\Psi^{(k)})}{p_h} - \sum_{j=1}^{m} \frac{\tau_g(x_j;\Psi^{(k)})}{p_h}, \quad \forall h = 1, 2, \ldots, (g-1) \qquad (8.15)$$

Applying arguments similar to those used for deriving Equation (8.10), we can show that Equation (8.15) leads to the following result:

$$p_h^{(k+1)} = \frac{\sum_{j=1}^{m} \tau_h(x_j;\Psi^{(k)})}{m}, \quad \forall h = 1, 2, \ldots, g \qquad (8.16)$$

On the other hand, by the first-order optimality condition with respect to θ_h, we have

$$0 = \frac{\partial E_{\Psi^{(k)}}[\log L_C(\Psi)|\mathbf{x}]}{\partial \theta_h}$$

$$= \frac{\partial \left\{ \sum_{j=1}^{m}\sum_{i=1}^{g} \tau_i(x_j;\Psi^{(k)})[\log p_i + \log f_i(x_j;\theta_i)] \right\}}{\partial \theta_h}$$

$$= \sum_{j=1}^{m} \frac{\partial \left\{ \sum_{i=1}^{g} \tau_i(x_j;\Psi^{(k)})[\log p_i + \log f_i(x_j;\theta_i)] \right\}}{\partial \theta_h}$$

$$= \sum_{j=1}^{m} \tau_h(x_j;\Psi^{(k)}) \frac{\partial \log f_h(x_j;\theta_h)}{\partial \theta_h} \qquad (8.17)$$

Note that Equations (8.13), (8.16), and (8.17) are the iterative forms of the MLE Equations (8.6), (8.10), and (8.11), respectively. The EM algorithm proceeds by executing the E and M steps alternatively until a certain stopping criterion is met. Dempster et al. (1977) have shown that the log-likelihood value is nondecreasing after a round of EM iteration. That is,

$$\log L(\psi^{(k+1)}) \geq \log L(\psi^{(k)})$$

If a corresponding upper bound exists, then the sequence of log-likelihood values converges to a limiting point. A continuity condition is needed to show that this limiting point is indeed a stationary value of the log-likelihood function (see Wu, 1983). The continuity condition also guarantees that if the parameters converge, they converge to a stationary point of the log-likelihood function. However, the convergence of the EM algorithm toward a local/global maximum requires stronger regularity conditions. Further details in this respect can be found in Wu (1983) and Mclachlan and Krishnan (1997).

In the special case of a univariate distribution with normal components, we have

$$f_h(x_j; \theta_h) = \frac{1}{\sigma_h \sqrt{2\pi}} \exp\left(-\frac{(x_j - \mu_h)^2}{2\sigma_h^2}\right)$$

Since

$$\log f_h(x_j; \theta_h) = -\frac{(x_j - \mu_h)^2}{2\sigma_h^2} - \log \sigma_h - \log \sqrt{2\pi}$$

we have the following derivatives:

$$\frac{\partial \log f_h(x_j; \theta_h)}{\partial \mu_h} = \frac{(x_j - \mu_h)}{\sigma_h^2}$$

$$\frac{\partial \log f_h(x_j; \theta_h)}{\partial \sigma_h} = \frac{(x_j - \mu_h)^2}{\sigma_h^3} - \frac{1}{\sigma_h}$$

By substituting the preceding two terms into Equation (8.17), we have

$$\sum_{j=1}^{m} \tau_h(x_j; \psi^{(k)}) \frac{(x_j - \mu_h)}{\sigma_h^2} = 0 \tag{8.18}$$

and

$$\sum_{j=1}^{m} \tau_h(x_j; \psi^{(k)}) \left[\frac{(x_j - \mu_h)^2}{\sigma_h^3} - \frac{1}{\sigma_h}\right] = 0 \tag{8.19}$$

Solving for μ_h in Equation (8.18) leads to

$$\mu_h^{(k+1)} = \frac{\sum_{j=1}^{m} \tau_h(x_j; \psi^{(k)}) x_j}{\sum_{j=1}^{m} \tau_h(x_j; \psi^{(k)})} \tag{8.20}$$

Equation (8.19) multiplied by σ_h^3 yields

$$\sum_{j=1}^{m} \tau_h(x_j; \Psi^{(k)})[(x_j - \mu_h)^2 - \sigma_h^2] = 0$$

which can be rewritten as

$$\sum_{j=1}^{m} \tau_h(x_j; \Psi^{(k)})(x_j - \mu_h)^2 = \sigma_h^2 \sum_{j=1}^{m} \tau_h(x_j; \Psi^{(k)})$$

Consequently,

$$(\sigma_h^{(k+1)})^2 = \frac{\sum_{j=1}^{m} \tau_h(x_j; \Psi^{(k)})(x_j - \mu_h^{(k+1)})^2}{\sum_{j=1}^{m} \tau_h(x_j; \Psi^{(k)})} \tag{8.21}$$

Note that we have added the superscript $(k+1)$ for μ_h and σ_h in Equations (8.20) and (8.21), respectively, to indicate the iterative updating scheme.

Next, we formally present the EM algorithm for fitting a normal mixture.

8.1.2.1 Algorithm Fit Normal Mixture

Step 0. (Initialization) Assume values of $x_j, j = 1, 2, \ldots, m$ and parameters of the normal mixture $\{p_h^{(0)}, \mu_h^{(0)}, \sigma_h^{(0)}, \forall h = 1, \ldots, g\}$. Let $k = 0$.

Step 1. Calculate the posterior probability $\tau_i(x_j; \Psi^{(k)})$ according to Equation (8.13):

$$\tau_i(x_j; \Psi^{(k)}) = \frac{p_i^{(k)} f_i(x_j; \mu_l^{(k)}, \sigma_i^{(k)})}{\sum_{h=1}^{g} p_h^{(k)} f_h(x_j; \mu_h^{(k)}, \sigma_h^{(k)})} \quad \forall i = 1, 2, \ldots, g; \quad j = 1, 2, \ldots, m$$

Step 2. Update mixture proportions $p_h^{(k+1)}$ by using Equation (8.16):

$$p_h^{(k+1)} = \frac{\sum_{j=1}^{m} \tau_h(x_j; \Psi^{(k)})}{m} \quad \forall h = 1, 2, \ldots, g$$

Step 3. Update component parameters $(\mu_h^{(k+1)}, \sigma_h^{(k+1)})$ according to Equations (8.20) and (8.21):

$$\mu_h^{(k+1)} = \frac{\sum_{j=1}^{m} \tau_h(x_j; \Psi^{(k)})x_j}{\sum_{j=1}^{m} \tau_h(x_j; \Psi^{(k)})} \quad \forall h = 1, 2, \ldots, g$$

$$(\sigma_h^{(k+1)})^2 = \frac{\sum_{j=1}^{m} \tau_h(x_j; \Psi^{(k)})(x_j - \mu_h^{(k+1)})^2}{\sum_{j=1}^{m} \tau_h(x_j; \Psi^{(k)})} \quad \forall h = 1, 2, \ldots, g$$

Step 4. Update the log-likelihood value $\log L(\Psi)$ according to Equation (8.3):

$$\log L(\Psi^{(k+1)}) = \sum_{j=1}^{m} \log \sum_{i=1}^{g} p_i^{(k+1)} f_i(x_j; \mu_l^{(k+1)}, \sigma_i^{(k+1)})$$

Step 5. If the prescribed criteria (which will be discussed later) are met, stop; otherwise, let $k = k + 1$, and go to step 1.

8.1.3 Related Issues in Model Fitting

8.1.3.1 Number of Components in a Mixture

Assessing the number of components in a mixture model is an important and difficult problem. Three approaches can be applied (Mclachlan and Peel, 2001). The first approach relies on the likelihood value. It either uses the likelihood ratio as the test statistic in a hypothesis test or tries to maximize the penalized likelihood value, in which the penalty increases with the number of components. The second approach is nonparametric. An example of such an approach is to statistically test the number of modes of the distribution and use that as the number of components. The third approach uses the method of moments to determine the number of components to include in the model.

Since we seek to approximate a *known* distribution in our application of mixture models, the problem of assessing the number of components is quite different. In our case, the number of components to use in the mixture depends on the desired accuracy of approximation and computational effort required. An increment in the number of components not only requires more computational effort during the MLE fitting of the mixture but also leads to a higher level of complexity in the application of the mixture (to estimate the expectation and variance of a performance measure for a given schedule). As we shall see later in this chapter, the latter grows quickly with an increment in the number of components. Hence it is desirable to use a small number of components for each processing time variable.

8.1.3.2 Starting Values and Stopping Criteria for the EM Algorithm

To initiate the EM algorithm, we need to specify a starting value of $\Psi^{(0)}$. One way to do this is to use random starting values. In the case of normal components, we can use the following approach (see McLachlan and Peel, 2001):

Calculate the sample mean and variance:

$$\bar{x} = \sum_{j=1}^{m} x_j$$

$$V = \sum_{j=1}^{m} (x_j - \bar{x})^2 / m \tag{8.22}$$

Generate the value of $\mu_i^{(0)}$ independently from $N(\bar{x}, V)$ for $I = 1, \ldots, g$.

Assign the following initial values to the component variances and the mixing proportions:

$$(\sigma_i^{(0)})^2 = V$$

$$p_i^{(0)} = 1/g \qquad (8.23)$$

With respect to stopping criteria, we can use either a prescribed number of EM iterations (to limit computational effort), or we can continue until the relative changes in the parameters and/or the log likelihood reach a preset tolerance value. For other stopping criteria, see Mclachlan and Peel (2001).

8.1.3.3 Adjusting Moments after the MLE Fitting of the Mixture

Since our application of mixture models pertains to estimating the first two moments of the performance measures for a given schedule, it is natural to require the mixture approximation of processing time to have values of the first two moments identical to those of the actual distribution. To that end, we make the following adjustments to the parameters obtained from the EM algorithm:

$$\mu_m = \sum_{i=1}^{g} p_i \mu_i$$

$$\mu_i' = \mu_o + \mu_i - \mu_m$$

$$(\sigma_i')^2 = \sigma_i^2 \frac{\sigma_o^2 + \mu_m^2 - \sum_{k=1}^{g} p_k \mu_k^2}{\sum_{k=1}^{g} p_k \sigma_k^2} \qquad (8.24)$$

Note that μ_m is expectation of the mixture before adjustment, μ_o and σ_o are the mean and standard deviation of the known processing time distribution, and μ_i' and σ_i' are the adjusted mean and standard deviation of the ith normal component of the mixture. Note that after the preceding adjustment, the expectation of the mixture model

$$E[X] = \sum_{i=1}^{g} p_i \mu_i' = \sum_{i=1}^{g} p_i (\mu_o + \mu_i - \mu_m)$$

$$= \mu_o + \sum_{i=1}^{g} p_i \mu_i - \mu_m = \mu_o \qquad (8.25)$$

[The first equality in Equation (8.25) follows by the law of total expectation.]

The variance of the mixture model is given by

$$
\begin{aligned}
\mathrm{var}[X] &= \sum_{i=1}^{g} p_i(\sigma_i')^2 + \sum_{i=1}^{g} p_i(\mu_i')^2 - \left(\sum_{i=1}^{g} p_i\mu_i'\right)^2 \\
&= \sum_{i=1}^{g} p_i\sigma_i^2 \frac{\sigma_o^2 + \mu_m^2 - \sum_{k=1}^{g} p_k\mu_k^2}{\sum_{k=1}^{g} p_k\sigma_k^2} + \sum_{i=1}^{g} p_i(\mu_i')^2 - \left(\sum_{i=1}^{g} p_i\mu_i'\right)^2 \\
&= \sigma_o^2 + \mu_m^2 - \sum_{k=1}^{g} p_k\mu_k^2 + \sum_{i=1}^{g} p_i(\mu_i')^2 - \left(\sum_{i=1}^{g} p_i\mu_i'\right)^2 \\
&= \sigma_o^2 + \left(\sum_{i=1}^{g} p_i\mu_i\right)^2 - \sum_{i=1}^{g} p_i\mu_i^2 + \sum_{i=1}^{g} p_i(\mu_i')^2 - \left(\sum_{i=1}^{g} p_i\mu_i'\right)^2
\end{aligned}
$$

$$(8.26)$$

[The first equality in Equation (8.26) follows by the law of total variance.]
Note that

$$
\begin{aligned}
\left(\sum_{i=1}^{g} p_i\mu_i\right)^2 - \left(\sum_{i=1}^{g} p_i\mu_i'\right)^2 &= \left(\sum_{i=1}^{g} p_i\mu_i - \sum_{i=1}^{g} p_i\mu_i'\right) \cdot \left(\sum_{i=1}^{g} p_i\mu_i + \sum_{i=1}^{g} p_i\mu_i'\right) \\
&= \left(\sum_{i=1}^{g} p_i(\mu_i - \mu_i')\right) \cdot \left(\sum_{i=1}^{g} p_i(\mu_i + \mu_i')\right) \\
&= \left(\sum_{i=1}^{g} p_i(\mu_m - \mu_o)\right) \cdot \left(\sum_{i=1}^{g} p_i(\mu_i + \mu_i')\right) \\
&= (\mu_m - \mu_o) \cdot \sum_{i=1}^{g} p_i(\mu_i + \mu_i')
\end{aligned}
$$

and

$$
\begin{aligned}
\sum_{i=1}^{g} p_i(\mu_i')^2 - \sum_{i=1}^{g} p_i\mu_i^2 &= \sum_{i=1}^{g} p_i[(\mu_i')^2 - \mu_i^2] \\
&= \sum_{i=1}^{g} p_i(\mu_i' - \mu_i)(\mu_i' + \mu_i) \\
&= \sum_{i=1}^{g} p_i(\mu_o - \mu_m)(\mu_i' + \mu_i) \\
&= (\mu_o - \mu_m) \sum_{i=1}^{g} p_i(\mu_i' + \mu_i)
\end{aligned}
$$

Hence Equation (8.26) reduces to

$$\text{var}[X] = \sigma_o^2 + (\mu_m - \mu_o) \cdot \sum_{i=1}^{g} p_i(\mu_i + \mu_i') + (\mu_o - \mu_m) \sum_{i=1}^{g} p_i(\mu_i' + \mu_i)$$

$$= \sigma_o^2$$

Next, we provide a numerical example to demonstrate the use of the mixture model to approximate a continuous distribution.

Example 8.1. Fitting a Mixture Model to a Continuous Uniform Distribution Consider a continuous uniform distribution of processing time

$$X \sim U(a, b)$$

where $a = 3$, $b = 6$. If we approximate X simply by a normal distribution, we have

$$X_G \sim N(\mu_G, \sigma_G^2)$$

The parameters of X_G can be determined by matching moments, which leads to

$$\mu_G = E[X_G] = E[X] = \frac{a+b}{2} = 4.5$$

$$\sigma_G^2 = \text{var}[X_G] = \text{var}[X] = \frac{(b-a)^2}{12} = 0.75$$

Next, consider the use of a normal mixture to approximate the same distribution. We first generate 2000 i.i.d. samples from the given uniform distribution. The sample mean and variance happen to be, respectively, $\bar{x} \approx 4.513$ and $V \approx 0.7489$. According to Equation (8.23), the initial values of component standard deviation and weights are

$$\sigma_i^{(0)} = \sqrt{V} \approx 0.8654$$

$$p_i^{(0)} = 1/5, \qquad \forall i = 1, \dots, 5$$

To get initial values of the component means, we simply take five random samples from normal distribution $N(4.513, 0.7489)$. The resulting parameters of the initial mixture are summarized below:

i	1	2	3	4	5
$p_i^{(0)}$	0.2	0.2	0.2	0.2	0.2
$\mu_i^{(0)}$	4.138	3.071	4.621	4.762	3.521
$\sigma_i^{(0)}$	0.8654	0.8654	0.8654	0.8654	0.8654

In the E step, we calculate the posterior probability $\tau_i(x_j; \Psi^{(0)})$, for $i = 1, 2, \ldots, g$ and $j = 1, 2, \ldots, m$. To illustrate, consider a particular data sample $x_j = 5.6559$. We first calculate the value

$$p_i^{(0)} f_i(x_j; \mu_i^{(0)}, \sigma_i^{(0)}), \qquad \forall i = 1, \ldots, 5$$

where $f_i(x_j; \mu_i^{(0)}, \sigma_i^{(0)})$ is the normal density function with mean $\mu_i^{(0)}$ and standard deviation $\sigma_i^{(0)}$. The results are

i	1	2	3	4	5
$p_i^{(0)} f_i(x_j; \mu_i^{(0)}, \sigma_i^{(0)})$	0.019815	0.0010661	0.045111	0.054058	0.0043916

The summation is

$$\sum_{i=1}^{5} p_i^{(0)} f_i(x_j; \mu_i^{(0)}, \sigma_i^{(0)})$$

$$= 0.019815 + 0.0010661 + 0.045111 + 0.054058 + 0.0043916$$

$$= 0.12444$$

Hence the posterior probability values are

i	1	2	3	4	5
$\tau_i(x_j; \Psi^{(0)})$	$\dfrac{0.019815}{0.12444}$ $= 0.15923$	$\dfrac{0.0010661}{0.12444}$ $= 0.0085675$	$\dfrac{0.045111}{0.12444}$ $= 0.36251$	$\dfrac{0.054058}{0.12444}$ $= 0.43441$	$\dfrac{0.0043916}{0.12444}$ $= 0.035291$

Having calculated the value of $\tau_i(x_j; \Psi^{(0)})$ for all i and j, we proceed to the M step. We update the weights by taking the average values of posterior probabilities across all 2000 sample points. That is,

$$p_i^{(1)} = \frac{\sum_{j=1}^{2000} \tau_i(x_j; \Psi^{(0)})}{2000}$$

The resulting values are

i	1	2	3	4	5
$p_i^{(1)}$	0.2128	0.1157	0.2518	0.2616	0.1585

The mean and variance of the mixing components are updated by recalculating sample mean and sample variance for each component and using the τ values as the recurring frequency of the sample values. More specifically,

$$\mu_i^{(1)} = \frac{\sum_{j=1}^{2000} \tau_i(x_j; \Psi^{(0)})x_j}{\sum_{j=1}^{2000} \tau_i(x_j; \Psi^{(0)})} \qquad \forall i = 1, \ldots, g$$

$$(\sigma_i^{(1)})^2 = \frac{\sum_{j=1}^{2000} \tau_i(x_j; \Psi^{(0)})(x_j - \mu_i^{(1)})^2}{\sum_{j=1}^{2000} \tau_i(x_j; \Psi^{(0)})} \qquad \forall i = 1, \ldots, g$$

This leads to

i	1	2	3	4	5
$\mu_i^{(1)}$	4.436	3.710	4.841	4.949	3.958
$\sigma_i^{(1)}$	0.7996	0.5857	0.7698	0.7439	0.7000

From here on, we repeat the E and M steps until the prescribed stopping criteria are met. The resulting distributions are shown in Figure 8.2. In the last diagram in Figure 8.2, we also have plotted the histogram of sample data, which closely matches the shape of the mixture.

Note that owing to random variation, the shape of the sample histogram is different from the p.d.f. of the continuous uniform distribution. Hence the final shape of the mixture does not follow the shape of the uniform distribution exactly. To reduce this type of bias, we consider an alternative approach as follows: Instead of generating the sample data randomly, we use the following data set:

$$\{x_j = 3 + 0.002j, \forall j = 0, 1, \ldots, 1500\}$$

Clearly, the histogram of this new data set matches the density function of continuous uniform more closely. The corresponding result from the EM algorithm is shown in Figure 8.3, where the target distribution ($U(3,6)$) is drawn in dashed lines. Note that this alternative approach gives a symmetric shape of the mixture, which is a desirable feature because the uniform distribution itself is symmetric.

Using the preceding alternative approach, we also approximate the same uniform distribution with mixtures of 10, 15, 20, and 25 components. The results are illustrated in Figure 8.4. Note that the p.d.f. of the mixture model converges to that of the uniform distribution with increments in the number of components.

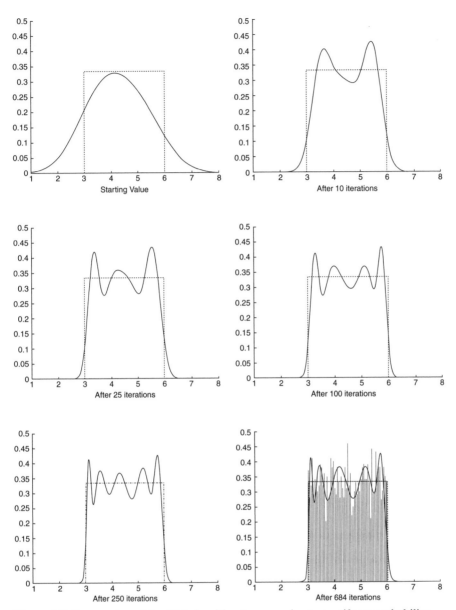

Figure 8.2. Progression of the EM algorithm to approximate a uniform probability density function.

8.2 Application of Mixture Models for Estimating the Moments of Various Performance Measures of a Schedule

In this section we demonstrate how mixture models can be used to derive approximations of the expectation and variance of performance measures for a given schedule. Note that a number of cases in Table 8.1 rely on the assumption of normal distribution. If simply a normal distribution is used to

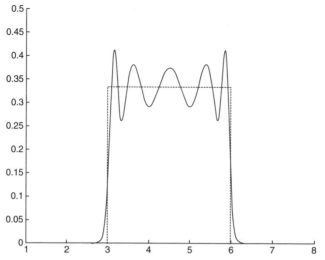

Figure 8.3. Probability density function of a normal mixture with alternative sample data (after 1684 EM iterations).

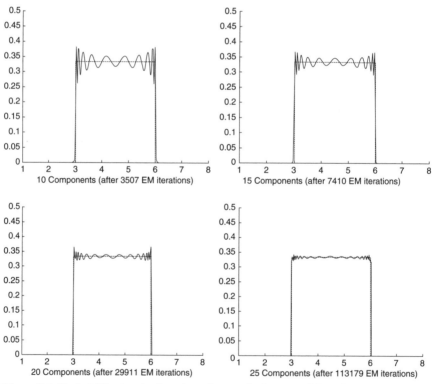

Figure 8.4. Probability density function of normal mixtures with various numbers of components.

approximate a general processing time distribution, an error inevitably will be introduced. Therefore, mixture models are useful for such cases. As shown in Section 8.1, we can approximate a general distribution quite effectively with a finite mixture of normal components. Besides, normal components naturally suit the methods employed in Chapters 4 through 7 to determine the expectation and variance of a performance measure. Therefore, the use of mixture models to estimate the moments of a performance measure is a promising way to approach the problem. We start our discussion with a simple example in Section 8.2.1 that addresses the tardiness measure. A unit penalty measure (to capture the number of tardy jobs) is similarly addressed in Section 8.2.2. Further applications of mixture models to determine the expectation and variance of various performance measures for a schedule in the single-machine environment are presented in Section 8.2.3. The use of mixture models for (more general) job-shop schedules is discussed in Section 8.2.4. Section 8.2.5 introduces the flow-shop and parallel-machine problems as special cases of the job-shop problem. Section 8.2.6 addresses issues related to computational efficiency of the proposed approach. Finally, in Section 8.2.7 we apply mixture models to estimate the mean and variance of the makespan for a stochastic activity network.

8.2.1 Estimating Expectation and Variance of Tardiness

Suppose that the completion time C of a given job follows a continuous uniform distribution $U(3,6)$ and that the due date of this job is $d = 5$. We shall illustrate determination of the expectation and variance of its tardiness by Clark's equations using the mixture model presented for uniform distribution in Section 8.1.

Let the p.d.f. $f_C(x)$ of the random variable C be represented by a mixture model as follows:

$$f_C(x) = \sum_{i=1}^{g} \pi_i f_{C_i}(x) \tag{8.27}$$

where π_i is the weight of its ith component. The ith component, C_i, is assumed to be normally distributed, that is,

$$C_i \sim N(\mu_i, \sigma_i^2)$$

To estimate the expectation and variance of tardiness $T = \max\{C - d, 0\}$, we claim that T is a mixture of random components T_i's that are defined as $T_i \equiv \max\{C_i - d, 0\}$, $i = 1, 2, \ldots, g$. The corresponding weights are π_i, $i = 1, 2, \ldots, g$. Note that $(C_i - d)$ is a normally distributed random variable. The deterministic value of 0 also can be viewed as a degenerate normal $N(0,0)$. Hence we can apply Clark's equations to obtain the following results for T_i [see Equations (4.17) and (4.18)]:

$$E[T_i] = v_i \Phi(\alpha_i) + \sigma_i \varphi(\alpha_i)$$

$$\text{var}[T_i] = (v_i^2 + \sigma_i^2)\Phi(\alpha_i) + v_i \sigma_i \varphi(\alpha_i) - E[T_i]^2,$$

where $v_i = \mu_i - d$ and $\alpha_i = v_i/\sigma_i$, for $i = 1, \ldots, g$. Subsequently, we have

$$E[T] = \sum_{i=1}^{g} \pi_i E[T_i],$$

$$\text{var}[T] = \sum_{i=1}^{g} \pi_i \text{var}[T_i] + \sum_{i=1}^{g} \pi_i E[T_i]^2 - E[T]^2 \qquad (8.28)$$

Using the mixture models obtained for uniform distribution in Example 8.1.1, we can calculate the expectation and variance of job tardiness as follows: Consider the five-component normal mixture fitted to the uniform distribution $U(3, 6)$. The parameters (obtained previously in Example 8.1.1) are

i	1	2	3	4	5
π_i	0.4687	0.1922	0.07343	0.07344	0.1922
μ_i	4.500	3.562	5.858	3.142	5.438
σ_i	0.4989	0.2465	0.08994	0.08996	0.2465

For the first component C_1, we calculate

$$v_1 = \mu_1 - d = 4.5 - 5 = -0.5$$

$$\alpha_1 = \frac{-0.5}{0.4989} = -1.002$$

Hence

$$E[T_1] = v_1 \Phi(\alpha_1) + \sigma_1 \varphi(\alpha_1)$$

$$= -0.5\Phi(-1.002) + 0.4989\varphi(-1.002)$$

$$= -0.5 \times 0.15813 + 0.4989 \times 0.24145$$

$$= 0.04139.$$

$$\text{var}[T_1] = (v_1^2 + \sigma_1^2)\Phi(\alpha_1) + v_1\sigma_1\varphi(\alpha_1) - E[T_1]^2$$

$$= ((-0.5)^2 + 0.4989^2)\Phi(-1.002) + (-0.5) \times 0.4989$$

$$\times \varphi(-1.002) - 0.04139^2$$

$$= 0.4987 \times 0.15813 + (-0.2493) \times 0.24145 - 0.0017132$$

$$= 0.01695.$$

With similar calculations for the other four components, we have the following:

i	1	2	3	4	5
π_i	0.4687	0.1922	0.07343	0.07344	0.1922
$E[T_i]$	0.04139	≈ 0	0.8576	≈ 0	0.4416
$\text{var}[T_i]$	0.01695	≈ 0	0.008090	≈ 0	0.05682

The overall expectation and variance of tardiness are given by

$$E[T] = \sum_{i=1}^{5} \pi_i E[T_i] = 0.16725$$

$$\text{var}[T] = \sum_{i=1}^{5} \pi_i \text{var}[T_i] + \sum_{i=1}^{5} \pi_i E[T_i]^2 - E[T]^2$$

$$= 0.019457 + 0.092290 - 0.027972$$

$$= 0.08377$$

We further apply mixture models with different numbers of components to approximate the expectation and variance of the same tardiness variable. The results are shown in Table 8.2. For comparative purposes, next we analytically calculate the true values of expectation and variance of tardiness.

Note that the tardiness value has a semicontinuous distribution with a mass of $\Pr[T = 0] = \Pr[C \leq 5] = 2/3$ at point zero. If conditioned on $\{T \geq 0\}$, the tardiness is uniformly distributed over the interval $[0, 1]$. Since the expectation and variance of $U(0, 1)$ are 1/2 and 1/12, respectively, we have

$$E[T] = E[T|T = 0]\Pr[T = 0] + E[T|T > 0]\Pr[T > 0]$$

$$= 0 \times \frac{2}{3} + \frac{1}{2} \times \frac{1}{3} = \frac{1}{6}$$

Table 8.2. *Estimating Expectation and Variance of Tardiness with Mixture Models*

Number of Components	Expectation	Variance
1	0.15153	0.11266
5	0.16725	0.08377
10	0.16659	0.08350
15	0.16659	0.08344
20	0.16666	0.08339
25	0.16665	0.08341
Analytical values	0.16667 (1/6)	0.08333 (1/12)

$$\text{var}[T] = \text{var}[T|T = 0]\Pr[T = 0] + \text{var}[T|T > 0]\Pr[T > 0]$$
$$+ E[T|T = 0]^2\Pr[T = 0] + E[T|T > 0]^2\Pr[T > 0] - E[T]^2$$
$$= 0 \times \frac{2}{3} + \frac{1}{12} \times \frac{1}{3} + 0 \times \frac{2}{3} + \frac{1}{4} \times \frac{1}{3} - \frac{1}{36}$$
$$= \frac{1}{12}$$

Clearly, a mixture model provides closer approximations of the true values of expectation and variance than that provided by the normal distribution (which is equivalent to the case of only one normal component). As the number of components increases, the approximations of expectation and variance tend to be more accurate, although the gap between an approximate value and the true value is not always decreasing monotonically.

8.2.2 Estimating Expectation and Variance of a Unit Penalty Function

Consider next the case of estimating the expectation and variance of a unit penalty function, which is defined as follows:

$$U = \begin{cases} 1, & \text{if } C > d \\ 0, & \text{if } C \le d \end{cases}$$

Note that, as also defined in Section 4.1, U captures whether or not a job is late.

Assuming the same mixture model of completion time as in Equation (8.27), we claim that U is a mixture of components U_i's with weights π_i, $i = 1, 2, \ldots, g$. The ith component U_i is determined as

$$U_i = \begin{cases} 1, & \text{if } C_i > d \\ 0, & \text{if } C_i \le d \end{cases}$$

Since C_i follows a normal distribution, we can apply our previous results from Section 4.3.3 directly. Thus we have

$$E[U_i] = 1 - \Pr[C_i \le d] = 1 - F_{C_i}(d),$$

By the law of total expectation, we further have

$$E[U] = \sum_{i=1}^{g} \pi_i E[U_i] \tag{8.29}$$

Hence

$$\text{var}[U] = E[U](1 - E[U]) \tag{8.30}$$

Consider the same job completion time distribution $(U(3,6))$ and due date $(d = 5)$ as in the preceding section. The parameter values of the mixture model for $C \sim U(3,6)$ consisting of five components were obtained in Example 8.1.1. We repeat them here for the sake of convenience:

i	1	2	3	4	5
π_i	0.4687	0.1922	0.07343	0.07344	0.1922
μ_i	4.500	3.562	5.858	3.142	5.438
σ_i	0.4989	0.2465	0.08994	0.08996	0.2465

For the first component C_1, we have

$$E[U_1] = 1 - F_{C_1}(d) = 1 - 0.84187 = 0.15813$$

where $F_{C_1}(\cdot)$ is the cumulative distribution function of a normal random variable with mean μ_1 and standard deviation σ_1. Proceeding similarly for $i = 2, \ldots, 5$, we have

i	1	2	3	4	5
$E[U_i]$	0.15813	0	1	0	0.96214

According to Equations (8.29) and (8.30), we obtain

$$E[U] = \sum_{i=1}^{5} \pi_i E[U_i] = 0.33248$$

$$\text{var}[U] = E[U](1 - E[U]) = 0.22194$$

Numerical results for various mixture models are provided in Table 8.3. The analytical values of $E[U]$ and $\text{var}[U]$ are as follows:

$$E[U] = \Pr[C > 5] = \frac{6-5}{6-3} = \frac{1}{3}$$

Table 8.3. *Estimating Expectation and Variance of a Unit Penalty Function with Mixture Models*

Number of Components	Expectation	Variance
1	0.28185	0.20241
5	0.33248	0.22194
10	0.33126	0.22153
15	0.33393	0.22242
20	0.33319	0.22217
25	0.33299	0.22211
Analytical values	0.33333 (1/3)	0.22222 (2/9)

and

$$\mathrm{var}[U] = E[U](1 - E[U]) = \frac{1}{3} \cdot \frac{2}{3} = \frac{2}{9}$$

Again, a mixture model provides more accurate approximations of the expectation and variance of U than that provided by a single normal distribution.

8.2.3 Single-Machine Problems

Next, we consider the use of mixture models to determine the expectation and variance of a performance measure of a schedule for the single-machine environment. We focus on the due-date-based performance measures, which necessitated the use of normal processing time distributions for analysis (see Table 8.1) and resulted in approximate solutions. Now we drop this assumption of normal processing time distributions and instead consider job processing times to have arbitrary continuous distributions that, as we showed earlier, can be approximated by mixtures of normal components. We assume the processing time distributions of the jobs to be independent.

To derive estimates of various performance measures, we first investigate the completion time distributions of jobs located at various positions in the schedule. Consider the job scheduled at the first position. Clearly, its completion time follows the distribution of its processing time (assuming the schedule start time to be zero). Next, consider the jth job ($j > 1$) of the schedule. Suppose that the following equation holds for $l = 1, 2, \ldots, j - 1$:

$$f_{C_{[l]}}(x) = \sum_{i=1}^{g_{[l]}^C} \pi_i^{[l]} f_{C_{[l]},i}(x) \tag{8.31}$$

That is, the completion time of each of the previous jobs, $l = 1, 2, \ldots, j - 1$, can be expressed as a mixture of $g_{[l]}^C$ normal components. We will show that Equation (8.31) holds for $l = j$ as well.

As noted earlier, the processing time of the jth job is approximated by a normal mixture:

$$f_{P_{[j]}}(x) = \sum_{i=1}^{g_{[j]}^P} p_i^{[j]} f_{P_{[j]},i}(x)$$

Since the completion time of the jth job can be calculated as

$$C_{[j]} = C_{[j-1]} + P_{[j]}$$

its distribution will be a mixture of $g_{[j]}^C = g_{[j-1]}^C \times g_{[j]}^P$ normal components, where each component (i) of $C_{[j]}$ corresponds to the summation of a component (i')

of $C_{[j-1]}$ and a component (i'') of $P_{[j]}$. That is,

$$C_{[j]}^i = C_{[j-1]}^{i'} + P_{[j]}^{i''}$$

We define

$$\pi_i^{[j]} = \pi_{i'}^{[j-1]} \times p_{i''}^{[j]}$$

where

$$i = i'' + (i' - 1) \cdot g_{[j]}^P \qquad (8.32)$$

Consequently, the p.d.f. of $C_{[j]}$ is

$$f_{C_{[j]}}(x) = \sum_{i=1}^{g_{[j]}^C} \pi_i^{[j]} f_{C_{[j]},i}(x)$$

Since both $C_{[j-1]}^{i'}$ and $P_{[j]}^{i''}$ have normal distributions, $C_{[j]}^i$ also follows a normal distribution whose parameters are easy to derive from those of $C_{[j-1]}^{i'}$ and $P_{[j]}^{i''}$. Therefore, it follows that $C_{[j]}$ is also a mixture of normal components.

This procedure can be applied iteratively to derive the completion time distributions for all the jobs in the schedule. This is shown in Figure 8.5.

8.2.3.1 Total Tardiness
Now that the completion time distributions are derived, we can use the approach introduced in Section 8.2.1 to estimate the expectation and variance of job tardiness $T_{[j]}$ for each job j. Note that the number of components of $T_{[j]}$ is the same as those of $C_{[j]}$.

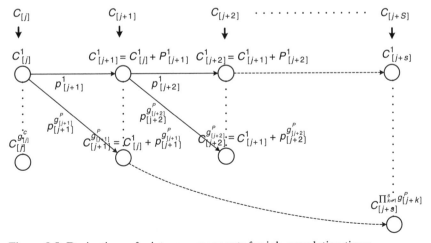

Figure 8.5. Derivations of mixture components for job completion times.

To derive the expectation and variance of total tardiness, we also need covariance terms $\text{cov}[T_{[j]}, T_{[j+s]}]$. Since $T_{[j]}$ and $T_{[j+s]}$ have, respectively, $g^C_{[j]}$ and $g^C_{[j+s]}$ components, there are $g^C_{[j]} \times g^C_{[j+s]}$ combinations to consider for their covariance. However, a useful property of these mixture components helps in reducing the amount of computation actually required. As shown in Figure 8.5, although $C_{[j+s]}$ has $g^C_{[j+s]} = g^C_{[j]} \cdot \prod_{k=1}^{s} g^P_{[j+k]}$ components, there are only $\prod_{k=1}^{s} g^P_{[j+k]}$ of these components that are correlated with the hth component of $C_{[j]}$. If we denote the index set of these components as $\gamma_h^{j,s}$, the covariance between $T_{[j]}$ and $T_{[j+s]}$ can be expressed as

$$\text{cov}[T_{[j]}, T_{[j+s]}] = \sum_{h=1}^{g^C_{[j]}} \sum_{k \in \gamma_h^{j,s}} \pi_k^{[j+s]} \text{cov}[T_{[j]}^h, T_{[j+s]}^k] \qquad (8.33)$$

where $T_{[j]}^h = \max\{C_{[j]}^h - d_{[j]}, 0\}$ and $T_{[j+s]}^k = \max\{C_{[j+s]}^k - d_{[j+s]}, 0\}$. Note that

$$C_{[j+s]}^k = C_{[j]}^h + \tilde{P}$$

where $\tilde{P}$ is the summation of normal components selected from $P_{[j+1]}$, $P_{[j+2]}, \ldots, P_{[j+s]}$ to generate the kth component of $C_{[j+s]}$. Note that $\tilde{P}$ is also normally distributed. Therefore, the calculation of $\text{cov}[T_{[j]}^h, T_{[j+s]}^k]$ is similar to that of Equation (4.20), which we have derived for normal distributions.

As an illustration of the material just presented, we provide a numerical example for the total tardiness measure. The problem data are presented in Table 8.4.

Consider the processing sequence 1–2–3. We approximate processing time of each job with a mixture of g normal components. The various values of g that are considered include $\{1, 5, 10, 15, 20\}$. Note that we can reuse the mixture models obtained in Example 8.1.1 by scaling the mixture parameters properly. Suppose that we have a g-component normal mixture for $U(a, b)$. To obtain the g-component normal mixture for $U(\tilde{a}, \tilde{b})$, we apply the following conversion:

$$\tilde{\pi}_i = \pi_i$$

Table 8.4. *Data for a Single-Machine Problem*

Job Number	1	2	3
Processing time	$U(20, 60)$	$U(10, 30)$	$U(70, 95)$
Due date	58	87	175

$$\tilde{\mu}_i = \frac{\tilde{b} - \tilde{a}}{b - a} \cdot \left(\mu_i - \frac{a + b}{2} \right) + \frac{\tilde{a} + \tilde{b}}{2}$$

$$\tilde{\sigma}_i = \frac{\tilde{b} - \tilde{a}}{b - a} \cdot \sigma_i, \qquad \forall i = 1, \ldots, g$$

where π_i, μ_i, and σ_i are, respectively, the weight, mean, and standard deviation of the ith component for the mixture of $U(a, b)$.

Consider $g = 5$, for example. Once again, we present below the parameter values that we determined for the mixture model of $U(3, 6)$ in Example 8.1.1:

i	1	2	3	4	5
π_i	0.4687	0.1922	0.07343	0.07344	0.1922
μ_i	4.500	3.562	5.858	3.142	5.438
σ_i	0.4989	0.2465	0.08994	0.08996	0.2465

After scaling, the mixture representation of $P_{[1]} \sim U(20, 60)$ is as follows:

i	1	2	3	4	5
π_i	0.4687	0.1922	0.07343	0.07344	0.1922
μ_i	40.001	27.496	58.101	21.899	52.504
σ_i	6.6514	3.2873	1.1992	1.1994	3.2869

The p.d.f. of the preceding mixture is depicted in Figure 8.6. Similarly, we can derive mixture models for $P_{[2]} \sim U(10, 30)$ and $P_{[3]} \sim U(70, 95)$.

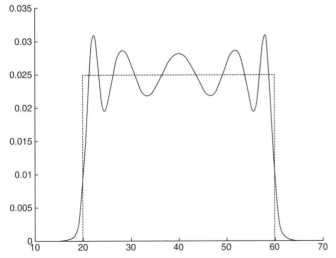

Figure 8.6. The five-component mixture model for $U(20, 60)$.

Next, we determine the mixture representations of $C_{[1]}$, $C_{[2]}$, and $C_{[3]}$, which have, respectively, 5, 25, and 125 components. Subsequently, the expectation and variance of $T_{[1]}$, $T_{[2]}$, and $T_{[3]}$ are obtained by applying the procedure presented in Section 8.2.1.

To illustrate the determination of covariance terms, consider, for example, $\text{cov}[T_{[1]}, T_{[2]}]$. By Equation (8.33),

$$\text{cov}[T_{[1]}, T_{[2]}] = \sum_{h=1}^{5} \sum_{k=5h-4}^{5h} \pi_k^{[2]} \text{cov}[T_{[1]}^h, T_{[2]}^k] \qquad (8.34)$$

Let $h = 2$ and $k = 8$. We have $C_{[2]}^i = C_{[1]}^{i'} + P_{[2]}^{i''}$, where $i = k = 8, i' = h = 2$.

According to Equation (8.32), $i'' = i - (i'' - 1)5 = 8 - (2-1)5 = 3$. Hence $C_{[2]}^8 = C_{[1]}^2 + P_{[2]}^3$. Note that $C_{[1]}^2 = P_{[1]}^2$. Consequently, $E[T_{[1]}^2 \cdot T_{[2]}^8]$ can be determined from Equation (4.20) by using $X = P_{[1]}^2$ and $Y = P_{[2]}^3$. According to the definition of covariance, we further have

$$\text{cov}[T_{[1]}^2, T_{[2]}^8] = E[T_{[1]}^2 \cdot T_{[2]}^8] - E[T_{[1]}^2] \cdot E[T_{[2]}^8]$$

Similar calculations can be carried out for other values of i and i' in Equation (8.34).

Finally, by Equations (4.14) and (4.15), we have

$$E\left[\sum_{j=1}^{n} T_{[j]}\right] = E[T_{[1]}] + E[T_{[2]}] + E[T_{[3]}]$$

$$\text{var}\left[\sum_{j=1}^{n} T_{[j]}\right] = \text{var}[T_{[1]}] + \text{var}[T_{[2]}] + \text{var}[T_{[3]}]$$

$$+ 2\text{cov}[T_{[1]}, T_{[2]}] + 2\text{cov}[T_{[1]}, T_{[3]}] + 2\text{cov}[T_{[2]}, T_{[3]}]$$

The numerical results are summarized in Table 8.5.

For reference, we also calculate the expectation and variance values analytically. The analysis involves lengthy convolution and multivariate integration. In order not to detract the reader, the details are presented in the Appendix.

Table 8.5. *Results for a Single-Machine Total Tardiness Problem*

Number of Components for $P_{[i]}$	$E[\sum_{j=1}^{n} T_j]$	$\text{var}[\sum_{j=1}^{n} T_j]$
1	0.45385	7.76302
5	0.08355	0.29087
10	0.07830	0.21981
15	0.07754	0.21546
20	0.07715	0.21333
25	0.07713	0.21308
Analytical values	$0.07646 \left(\dfrac{367}{4800}\right)$	$0.21042 \left(\dfrac{24240667}{115200000}\right)$

The corresponding results are shown in the last row of Table 8.5. Note that when a normal mixture model has only one component, it is equivalent to a normal distribution. Therefore, in that case we can use the approach presented in Section 4.3.1 to estimate the expectation and variance. However, these approximate values are less accurate because a uniform distribution cannot be approximated well by a normal distribution. When we increase the number of components in the mixture model, we obtain a more accurate approximation of the uniform distribution. Hence the values of expectation and variance approach their true values.

To explain this point pictorially, we illustrate the p.d.f.'s of completion times in Figure 8.7. The p.d.f.'s of $C_{[1]}$, $C_{[2]}$, and $C_{[3]}$ are drawn, respectively, in

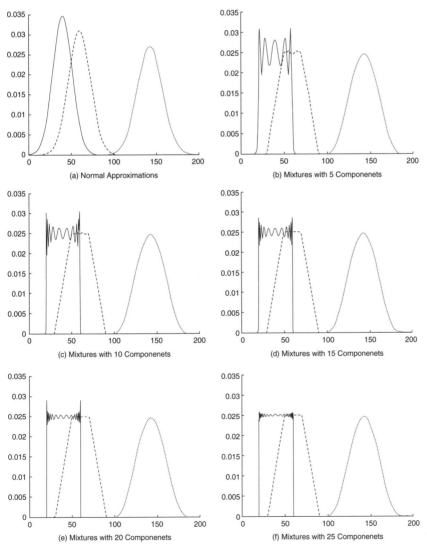

Figure 8.7. Completion time distributions obtained using mixture models of various numbers of components.

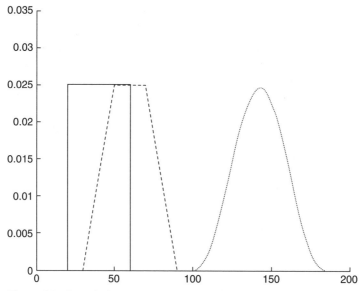

Figure 8.8. Completion time distributions (analytical results).

solid, dashed, and dotted lines. We also have provided the analytical p.d.f.'s of $C_{[1]}$, $C_{[2]}$, and $C_{[3]}$ in Figure 8.8. Note that the approximations achieved by mixture models (e.g., see Figure 8.7*b–f*) are closer to the true completion time distributions than those achieved by the normal approximations (see Figure 8.7*a*).

8.2.3.2 Total Weighted Tardiness

Having calculated $E[T_{[j]}]$ and $\text{cov}[T_{[i]}, T_{[j]}]$ earlier [see Equations (8.28) and (8.33)], we can use Equation (4.21) from Section 4.3.2 to determine the expectation and variance of total weighted tardiness.

As a numerical example, consider the single-machine problem presented in Table 8.4. Assume the weights for jobs 1, 2, and 3 to be 3, 2, and 1, respectively. Then we have

$$E\left[\sum_{j=1}^{n} w_{[j]} T_{[j]}\right] = 3 \times E[T_{[1]}] + 2 \times E[T_{[2]}] + E[T_{[3]}],$$

$$\text{var}\left[\sum_{j=1}^{n} w_{[j]} T_{[j]}\right] = 3^2 \times \text{var}[T_{[1]}] + 2^2 \times \text{var}[T_{[2]}] + \text{var}[T_{[3]}]$$

$$+ 2 \times 3 \times 2 \times \text{cov}[T_{[1]}, T_{[2]}] + 2 \times 3 \times 1 \times \text{cov}[T_{[1]}, T_{[3]}]$$

$$+ 2 \times 2 \times 1 \times \text{cov}[T_{[2]}, T_{[3]}].$$

The results are presented in Table 8.6. The analytical values are obtained by using the preceding equations and the analytical values of expectation,

Table 8.6. *Results for a Single-Machine Total Weighted Tardiness Problem*

Number of Components for $P_{[i]}$	$E[\sum_{j=1}^{n} w_j T_j]$	$\text{var}[\sum_{j=1}^{n} w_j T_j]$
1	1.13041	42.19264
5	0.19982	1.36024
10	0.18702	0.93720
15	0.18496	0.91142
20	0.18393	0.89942
25	0.18385	0.89768
Analytical values	$0.18208 \left(\dfrac{437}{2400} \right)$	$0.88356 \left(\dfrac{25446643}{28800000} \right)$

variance, and covariance derived in Section A.1 of the Appendix. Notice the closeness of the values obtained using the mixture models to the analytical values.

8.2.3.3 Total Number of Tardy Jobs

According to our previous results in Section 4.3.3, the expectation and variance of the total number of tardy jobs can be determined by Equations (4.24) and (4.25), which depend on the following three terms:

$$\mu_{U_{[j]}} = E[U_{[j]}]$$
$$\sigma^2_{U_{[j]}} = \text{var}[U_{[j]}]$$

and

$$\text{cov}[U_{[j]}, U_{[j+s]}] = E[U_{[j]} \cdot U_{[j+s]}] - E[U_{[j]}] \cdot E[U_{[j+s]}] \tag{8.35}$$

Calculation of the first two terms ($E[U_{[j]}]$ and $\text{var}[U_{[j]}]$) was discussed in Section 8.2.2. We need to further determine the value of $E[U_{[j]} \cdot U_{[j+s]}]$ using mixture models. Note that the completion times $C_{[j]}$ and $C_{[j+s]}$ have, respectively, $g^C_{[j]}$ and $g^C_{[j+s]}$ components. Similar to the discussion for the total tardiness case, there are only $\prod_{k=1}^{s} g^P_{[j+k]}$ components in $C_{[j+s]}$ that are correlated with a particular component (h) of $C_{[j]}$. Denoting the index set of these components by the notation of $\gamma_h^{j,s}$, as in Section 8.2.3.1, we have

$$E[U_{[j]} \cdot U_{[j+s]}] = \sum_{h=1}^{g^C_{[j]}} \sum_{k \in \gamma_h^{j,s}} \pi_k^{[j+s]} E[U_{[j]}^h \cdot U_{[j+s]}^k] \tag{8.36}$$

where $U_{[j]}^h$ corresponds to $C_{[j]}^h$ (the hth component of $C_{[j]}$), and $U_{[j+s]}^k$ corresponds to $C_{[j+s]}^k$ (the kth component of $C_{[j+s]}$).

Note that

$$C_{[j+s]}^k = C_{[j]}^h + \tilde{P}$$

where $\tilde{P}$ is the summation of normal components selected from $P_{[j+1]}$, $P_{[j+2]}, \ldots, P_{[j+s]}$ to generate the (i')th component of $C_{[j+s]}$. Since $\tilde{P}$ also has a normal distribution, calculation of $E[U_{[j]}^h, U_{[j+s]}^k]$ is similar to that of Equation (4.28), which we derived in Section 4.3.3.

Again, to illustrate, consider the single-machine problem presented in Table 8.4. We first derive mixture models for $C_{[1]}$, $C_{[2]}$, and $C_{[3]}$, as we did in Section 8.2.3.1. The expectation and variance of $U_{[1]}$, $U_{[2]}$, and $U_{[3]}$ are calculated by using the approach presented in Section 8.2.2. Next, we determine the covariance terms according to Equation (8.35). First, note that

$$\mathrm{cov}[U_{[1]}, U_{[2]}] = \sum_{h=1}^{5} \sum_{k=5h-4}^{5h} \mathrm{cov}[U_{[1]}^h, U_{[2]}^k] \qquad (8.37)$$

Let $h = 2$ and $k = 8$. We have $C_{[2]}^i = C_{[1]}^{i'} + P_{[2]}^{i''}$, where $i = k = 8$, $i' = h = 2$. According to Equation (8.32), $i'' = i - (i' - 1)5 = 8 - (2 - 1)5 = 3$.

Hence $C_{[2]}^8 = C_{[1]}^2 + P_{[2]}^3$. Note that $C_{[1]}^2 = P_{[1]}^2$. Consequently, $E[U_{[1]}^2 \cdot U_{[2]}^8]$ can be determined from Equation (4.28) by using $X = P_{[1]}^2$ and $Y = P_{[2]}^3$. According to the definition of covariance, we further have

$$\mathrm{cov}[U_{[1]}^2, U_{[2]}^8] = E[U_{[1]}^2 \cdot U_{[2]}^8] - E[U_{[1]}^2] \cdot E[U_{[2]}^8]$$

Similar calculations can be performed for other values of i and i' in Equation (8.37).

Finally, by Equations (4.24) and (4.25), we have

$$E\left[\sum_{j=1}^{n} U_{[j]}\right] = E[U_{[1]}] + E[U_{[2]}] + E[U_{[3]}]$$

$$\mathrm{var}\left[\sum_{j=1}^{n} U_{[j]}\right] = \mathrm{var}[U_{[1]}] + \mathrm{var}[U_{[2]}] + \mathrm{var}[U_{[3]}]$$

$$+ 2\mathrm{cov}[U_{[1]}, U_{[2]}] + 2\mathrm{cov}[U_{[1]}, U_{[3]}] + 2\mathrm{cov}[U_{[2]}, U_{[3]}]$$

The numerical results are presented in Table 8.7. The derivation of the analytical results (in the last row of the table) is presented in Section A.2 of the Appendix. Clearly, the mixture model provides more accurate results than those obtained by using a normal approximation.

8.2.3.4 Total Weighted Number of Tardy Jobs

In previous sections we have shown how to determine the values of $E[U_{[j]}]$, $\mathrm{var}[U_{[j]}]$, and $\mathrm{cov}[U_{[i]}, U_{[j]}]$ [see Equations (8.29) and (8.30) in Section 8.2.2 and Equation (8.35) in Section 8.2.3.3]. We can apply Equations (4.29) and (4.30) from Section 4.3.4 to determine the expectation and variance of the

Table 8.7. *Results Regarding Total Number of Tardy Jobs for a Single-Machine Problem*

Number of Components for $P_{[i]}$	$E[\sum_{j=1}^{n} U_j]$	$\mathrm{var}[\sum_{j=1}^{n} U_j]$
1	0.09176	0.15279
5	0.06437	0.08299
10	0.06529	0.08329
15	0.06482	0.08268
20	0.06397	0.08170
25	0.06428	0.08203
Analytical	$0.06396 \left(\dfrac{307}{4800} \right)$	$0.08160 \left(\dfrac{1880087}{23040000} \right)$

Table 8.8. *Results Regarding Total Weighted Number of Tardy Jobs for a Single-Machine Problem*

Number of Components for $P_{[i]}$	$E[\sum_{j=1}^{n} w_j U_j]$	$\mathrm{var}[\sum_{j=1}^{n} w_j U_j]$
1	0.22904	0.85949
5	0.17016	0.54835
10	0.17345	0.55607
15	0.17207	0.55155
20	0.16956	0.54388
25	0.17047	0.54662
Analytical	$0.16958 \left(\dfrac{407}{2400} \right)$	$0.54367 \left(\dfrac{3131567}{5760000} \right)$

total weighted number of tardy jobs. Note that $\mu_{U[j]} \equiv E[U_{[j]}]$ and $\sigma_{U[ij]} \equiv \mathrm{cov}[U_{[i]}, U_{[j]}]$.

We demonstrate this approach by applying it to the example presented in Section 8.2.3.1 (Table 8.4). Assume that the weights for jobs 1, 2, and 3 are 3, 2, and 1, respectively. Then we have

$$E\left[\sum_{j=1}^{n} w_{[j]} U_{[j]} \right] = 3 \times E[U_{[1]}] + 2 \times E[U_{[2]}] + E[U_{[3]}]$$

$$\mathrm{var}\left[\sum_{j=1}^{n} w_{[j]} U_{[j]} \right] = 3^2 \times \mathrm{var}[U_{[1]}] + 2^2 \times \mathrm{var}[U_{[2]}] + \mathrm{var}[U_{[3]}]$$

$$+ 2 \times 3 \times 2 \times \mathrm{cov}[U_{[1]}, U_{[2]}] + 2 \times 3 \times 1 \times \mathrm{cov}[U_{[1]}, U_{[3]}]$$

$$+ 2 \times 2 \times 1 \times \mathrm{cov}[U_{[2]}, U_{[3]}]$$

The numerical results are presented in Table 8.8. The analytical values are obtained by using the preceding equations and the analytical values of expectation, variance, and covariance derived in Section A.2 of the Appendix.

Again, the expectation and variance values obtained using mixture models are much closer to the actual values than those obtained by using a normal distribution.

8.2.3.5 Maximum Lateness

Recall that all job completion times are represented as normal mixtures. The completion time of job $[j]$ has the following p.d.f.:

$$f_{C_{[j]}}(x) = \sum_{i=1}^{g_{[j]}^C} \pi_i^{[j]} f_{C_{[j]},i}(x), \quad \text{for } j = 1, \ldots, n$$

where n is the number of jobs. For any job completion time $C_{[j]}, j = 1, \ldots, n$, the corresponding lateness $L_{[j]} = C_{[j]} - d_{[j]}$ also can be represented as a normal mixture:

$$f_{L_{[j]}}(x) = \sum_{i=1}^{g_{[j]}^C} \pi_i^{[j]} f_{L_{[j]},i}(x)$$

where $L_{[j]}^i = C_{[j]}^i - d_{[j]}, i = 1, \ldots, g_{[j]}^C$. According to Figure 8.5,

$$g_{[n]}^C = \prod_{j=1}^n g_{[j]}^P$$

This is also the number of possible combinations that we need to consider so that a normal component is selected from each of the processing time variables. Therefore, the maximum lateness $L_{\max}$ has the same number of components. That is,

$$g_{L_{\max}} = g_{[n]}^C$$

Furthermore, the weights for $L_{\max}$ are the same as those for $C_{[n]}$, which are $\pi_i^{[n]}, i = 1, \ldots, g_{[n]}^C$.

Now consider the ith component of this mixture:

$$L_{\max}^i = \max \left\{ (P_{[1]}^{i_1} - d_{[1]}), (P_{[1]}^{i_1} + P_{[2]}^{i_2} - d_{[2]}), \ldots, \left(\sum_{k=1}^n P_{[k]}^{i_k} - d_{[n]} \right) \right\} \quad (8.38)$$

where $i_1, i_2, \ldots, i_n$ are the indices of the components selected from the mixtures of $P_{[1]}, P_{[2]}, \ldots, P_{[n]}$ to generate the ith component of $C_{[n]}$. If we ignore the superscripts, the preceding expression is essentially the same as Equation (4.31). Therefore, we can apply the same approach as in Section 4.3.6 to estimate the values of $E[L_{\max}^i]$ and $\text{var}[L_{\max}^i]$.

Finally, we apply the law of total expectation and total variation on the mixture of $L_{\max}$:

$$f_{L_{\max}}(x) = \sum_{i=1}^{g_{[n]}^C} \pi_i^{[n]} f_{L_{\max},i}(x)$$

This leads to

$$E[L_{max}] = \sum_{i=1}^{g_{[n]}^C} \pi_i^{[n]} E[L_{max}^i];$$

$$var[L_{max}] = \sum_{i=1}^{g_{[n]}^C} \pi_i^{[n]} var[L_{max}^i] + \sum_{i=1}^{g_{[n]}^C} \pi_i^{[n]} E[L_{max}^i]^2 - E[L_{max}]^2$$

To illustrate, consider the sample problem depicted in Table 8.4 and the processing sequence of 1–2–3. Suppose that we use five-component mixtures to represent processing times. Accordingly, the maximum lateness has $g_{[1]}^P \times g_{[2]}^P \times g_{[3]}^P = 5 \times 5 \times 5 = 125$ components. The indices in Equation (8.38) can take the following values:

i	i_1	i_2	i_3
1	1	1	1
2	1	1	2
⋮	⋮	⋮	⋮
5	1	1	5
6	1	2	1
7	1	2	2
⋮	⋮	⋮	⋮
10	1	2	5
⋮	⋮	⋮	⋮
21	1	5	1
22	1	5	2
⋮	⋮	⋮	⋮
25	1	5	5
⋮	⋮	⋮	⋮
125	5	5	5

For each combination of $\langle i_1, i_2, i_3 \rangle$, we can calculate the expectation and variance of L_{max}^i by the approach presented in Section 4.3.6. Finally,

$$E[L_{max}] = \sum_{i=1}^{125} \pi_i^{[n]} E[L_{max}^i]$$

Table 8.9. *Results for Single-Machine Maximum Lateness Problem*

Number of Components for $P_{[i]}$	$E[\max\{L_j\}]$	$\text{var}[\max\{L_j\}]$
1	−17.64986	135.45061
5	−17.80827	134.01793
10	−17.81084	133.99640
15	−17.81113	133.99514
20	−17.81135	133.99409
25	−17.81132	133.99415
Analytical	$-17.81167 \left(-\dfrac{10687}{600}\right)$	$133.99236 \left(\dfrac{48237251}{360000}\right)$

$$\text{var}[L_{\max}] = \sum_{i=1}^{125} \pi_i^{[n]} \text{var}[L_{\max}^i] + \sum_{i=1}^{125} \pi_i^{[n]} E[L_{\max}^i]^2 - E[L_{\max}]^2$$

where $\pi_i^{[n]} = p_{i_1}^{[1]} \times p_{i_2}^{[2]} \times p_{i_3}^{[3]}$.

Numerical results obtained with various numbers of mixing components are presented in Table 8.9. The derivation of analytical values is provided in Section A.3 of the Appendix. Clearly, the mixture model gives more accurate approximations of expectation and variance for the maximum lateness measure.

8.2.4 Job-Shop Problems

Next we consider a schedule for a job-shop environment. An operation of a job t in the schedule is to be performed on a machine i in the jth position after completion of its previous operation performed on a different machine as well as after completion of the operation performed in the $(j-1)$th position on machine i. In order to determine the completion time distribution of the operation in the jth position on machine i, designated as $[i,j]$, we need to have information about (1) its processing time $P_{[i,j]}$ and (2) the joint distribution of the completion time of the previous operation of job t, which we denote as $C_{[p,q]}$, and the completion time of the operation scheduled in the $(j-1)$th position on machine i, $C_{[i,j-1]}$. In our previous discussion in Chapter 6, we assumed $P_{[i,j]}$ to follow a normal distribution and $C_{[p,q]}$ and $C_{[i,j-1]}$ to be jointly distributed according to a two-dimensional normal distribution. In the context of mixture models, we have assumed that all processing times are represented as normal mixtures. Accordingly, we will describe the joint distribution of $(C_{[p,q]}, C_{[i,j-1]})$ by a mixture of two-dimensional normal components. More generally, we use a mixture of multidimensional normal components to describe the joint distribution of the completion times of the "latest operations" $O\lambda$, which are

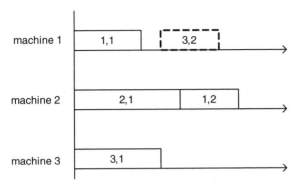

Figure 8.9. Partial sample job-shop schedule.

defined as follows:

$$O_\lambda = O_{\lambda M} \cup O_{\lambda J}$$

where

$O_{\lambda M} = \{\text{last operation scheduled on machine } i, i = 1, 2, \ldots, m\}$

$O_{\lambda J} = \{\text{last scheduled operation of job } t, t = 1, 2, \ldots, n\}$

For example, consider a job shop with three machines and three jobs. The current partial schedule is shown in Figure 8.9. According to our definition,

$$O_{\lambda M} = \{(1,1), (1,2), (3,1)\}, \qquad O_{\lambda J} = \{(1,2), (2,1), (3,1)\}$$
$$O_\lambda = O_{\lambda M} \cup O_{\lambda J} = \{(1,1), (1,2), (3,1), (2,1)\}$$

Note that the notation (x, y) represents the yth operation of job x.

If the next operation to be scheduled is the second operation of job 3, then, after it is scheduled, the updated values are

$$\tilde{O}_{\lambda M} = \{(3,2), (1,2), (3,1)\}$$
$$\tilde{O}_{\lambda J} = \{(1,2), (2,1), (3,2)\}$$
$$\tilde{O}_\lambda = \tilde{O}_{\lambda M} \cup \tilde{O}_{\lambda J} = \{(3,2), (1,2), (3,1), (2,1)\}$$

We denote the set of completion times of the operations in O_λ by C_λ, which is an $|O_\lambda|$-dimensional vector. The joint distribution of C_λ is to be updated whenever a new operation is added to the partial schedule. Note that when the schedule is completed, O_λ consists of the last operations of all the jobs and the last completed operations on all the machines. Having determined the distribution of C_λ, it would be easy to obtain the expectation and variance of various performance measures. For example, suppose that we focus on one component of the mixture that is used to represent C_λ. The corresponding makespan is the maximum of the elements in vector C_λ. Hence we can apply the same approach as Equation (6.7) to determine the expectation and variance of the makespan with respect to each component of C_λ. By applying the law of total expectation and the law of total variance, we can further derive overall expectation and variance of the makespan from their component-wise values.

Next, we present the updating scheme for the joint distribution of C_λ. As an induction hypothesis, we assume that the joint distribution can be expressed as

$$f_\lambda(\mathbf{c}) = \sum_{k=1}^{g^\lambda} \pi_k^\lambda f_{\lambda,k}(\mathbf{c}; \boldsymbol{\mu}_k^\lambda, \mathbf{S}_k^\lambda)$$

where $\mathbf{c}$ is a realization of C_λ; $\boldsymbol{\mu}_k^\lambda$ and $\mathbf{S}_k^\lambda$ are, respectively, the expectation vector and covariance matrix for the kth component (which is multivariate normal). We also assume the processing time of the next operation ($P_{[i,j]}$) to have a mixture normal distribution:

$$f_P(x) = \sum_{k=1}^{g^P} p_k f_k(x, \mu_k^P, (\sigma_k^P)^2)$$

After the addition of this operation, we obtain the updated set of $\tilde{O}_\lambda$. We approximate the joint distribution of its completion times ($\tilde{C}_\lambda$) by

$$\tilde{f}_\lambda(\tilde{\mathbf{c}}) = \sum_{k=1}^{\tilde{g}^\lambda} \tilde{\pi}_k^\lambda \tilde{f}_{\lambda,k}(\tilde{\mathbf{c}}; \tilde{\boldsymbol{\mu}}_k^\lambda, \mathbf{S}_k^\lambda)$$

where $\tilde{g}^\lambda = g^\lambda \times g^P$ and $\tilde{\pi}_k^\lambda = \pi_{k'}^\lambda \times p_{k''}$, assuming that the kth normal component of $\tilde{f}_\lambda$ is obtained by the combination of the (k')th component of f_λ and the (k'')th component of f_P. The parameters for the kth component of $\tilde{f}_\lambda$ (i.e., $\tilde{\boldsymbol{\mu}}_k^\lambda$ and $\tilde{\mathbf{S}}_k^\lambda$) can be determined accordingly as follows: First, note that

$$\tilde{O}_\lambda = (O_\lambda - \{O_{[p,q]}, O_{[i,j-1]}\}) \cup \{O_{[i,j]}\}$$

where $O_{[p,q]}$ is the previous operation of the same job to which operation $O_{[i,j]}$ belongs, and $O_{[i,j-1]}$ is the last operation processed on machine i before the insertion of $O_{[i,j]}$. If $O_{[i,j]}$ is the first operation of a job, then $O_{[p,q]}$ represents the (deterministic) release time of that job. If $O_{[i,j]}$ is the first operation scheduled on machine i (i.e., $j = 1$), then $O_{[i,j-1]}$ represents the (deterministic) ready time of machine i. Since most elements of O_λ are kept in set $\tilde{O}_\lambda$, determination of the updated expectation vector $\tilde{\boldsymbol{\mu}}_k^\lambda$ requires only calculation of the expectation of $C_{[i,j],k}$, which denotes the completion time of $O_{[i,j]}$ with respect to the kth component of $\tilde{f}_\lambda$. Similarly, to update the covariance matrix $\tilde{\mathbf{S}}_k^\lambda$, we need only the variance of $C_{[i,j],k}$ and its covariance with the other elements in $\tilde{C}_\lambda$. Note that

$$C_{[i,j],k} = \max\{C_{[i,j-1],k'}, C_{[p,q],k'}\} + P_{[i,j],k''} \tag{8.39}$$

The expectation and variance of $C_{[i,j],k}$ can be determined by using the approach outlined for Equation (6.2) in Chapter 6. The calculation of covariance terms is similar to that of Equation (6.6). Note that the expectations of $C_{[p,q]}$ and $C_{[i,j-1]}$ are elements of the expectation vector $\boldsymbol{\mu}_{k'}^\lambda$, and their variance and covariance are elements of the covariance matrix $\mathbf{S}_{k'}^\lambda$.

The preceding discussion can be summarized in the following algorithm to apply the mixture model for determining the expectation and variation of a given job-shop schedule.

8.2.4.1 Algorithm MixtureJobShop

Ω = set of unscheduled operations (its initial value consists of all the operations)

O_λ = set of latest operations (its initial value consists of job release times and machine ready times)

C_λ = completion times of operations in O_λ

ξ = a given set of operation sequencing decisions for all the machines

Step 1. Select operation $O_{[i,j]}$ from Ω such that $O_{[i,j]}$ is ready to be processed according to ξ. If no such operation exists, go to step 4.

Step 2. Let $\tilde{O}_\lambda = (O_\lambda - \{O_{[p,q]}, O_{[i,j-1]}\}) \cup \{O_{[i,j]}\}$. Calculate the mixture distribution of $\tilde{C}_\lambda$ according to the distributions of $\mathbf{C}_\lambda$ and $P_{[i,j]}$.

Step 3. Let $\Omega = \Omega - \{O_{[i,j]}\}$, $O_\lambda = \tilde{O}_\lambda$. Go to step 1.

Step 4. For $k = 1, \ldots, g^\lambda$, calculate the expectation and variance of performance measure Z according to the kth component of $\mathbf{C}_\lambda$. Record the results as $E[Z_k]$ and var$[Z_k]$, respectively.

Step 5. Calculate the expectation and variance of the performance measure as follows:

$$E[Z] = \sum_{k=1}^{g^\lambda} \pi_k E[Z_k],$$

$$\text{var}[Z] = \sum_{k=1}^{g^\lambda} \pi_k \text{var}[Z_k] + \sum_{k=1}^{g^\lambda} \pi_k \pi_k E[Z_k]^2 - E[Z]^2$$

Next, we illustrate this algorithm by using a numerical example. Consider a job-shop makespan problem with three jobs and two machines. The data are presented in Tables 8.10, 8.11, and 8.12.

Table 8.10. *Processing Time Distributions for the Operations of the Jobs*

	Job 1	Job 2	Job 3
Machine 1	$U[3,6]$	$U[2,7]$	$U[9,17]$
Machine 2	$U[6,7]$	$U[5,11]$	$U[1,8]$

Table 8.11. *Routings of Jobs*

Jobs	Routing of a Job on Machines	
Job 1	2	1
Job 2	1	2
Job 3	1	2

Table 8.12. *Machine Processing Sequence*

Machines	Sequence of Jobs on a Machine		
Machine 1	2	3	1
Machine 2	1	2	3

We approximate processing time of each operation with a mixture of five normal components. The initial set of "latest operations" is determined as follows:

$$O_{\lambda M} = \{O_{[1,0]}, O_{[2,0]}\}$$

$$O_{\lambda J} = \{(1,0), (2,0), (3,0)\}$$

$$O_{\lambda} = O_{\lambda M} \cup O_{\lambda J} = \{O_{[1,0]}, O_{[2,0]}, (1,0), (2,0), (3,0)\}$$

where $O_{[i,0]}$ is the ready time of machine i, $i = 1, 2$, and $(t, 0)$ is the release time of job t, $t = 1, 2, 3$. All these "operations" are assumed to be completed at deterministic time zero. Accordingly, the set of unscheduled jobs is

$$\Omega = \{(1,1), (1,2), (2,1), (2,2), (3,1), (3,2)\}$$

The mixture distribution of $\mathbf{C}_\lambda$ has only one component, which is degenerate multivariate normal.

Consider operation $(1, 1)$ to be scheduled next on machine 2 (i.e., $O_{[i,j]} = O_{[2,1]} = (1,1)$). Note that $O_{[p,q]} = (1,0)$, and $O_{[i,j-1]} = O_{[2,0]}$. The updated set of "latest operations" is

$$\tilde{O}_\lambda = (O_\lambda - \{O_{[p,q]}, O_{[i,j-1]}\}) \cup \{O_{[i,j]}\}$$

$$= \{O_{[1,0]}, O_{[2,1]}, (2,0), (3,0)\}$$

Since C_λ has one component and $P_{[i,j]}$ has five components, the updated random vector $\tilde{C}_\lambda$ can be represented by a five-component multivariate normal mixture, with its kth component being determined by the $(k' = 1)$th component of C_λ and the $(k'' = k)$th component of $P_{[i,j]}$.

Note that, in this case, we have

$$C_{[i,j],k} = \max\{C_{[i,j-1],k'}, C_{[p,q],k'}\} + P_{[i,j],k''}$$
$$= \max\{0, 0\} + P_{[i,j],k''}$$
$$= 0 + P_{[i,j],k''}$$
$$= P_{[i,j],k''}$$

and

$$\mathrm{cov}(C_{[i,j],k}, C_{[u,v],k}) = 0 \qquad \forall [u, v] \in \tilde{O}_\lambda \text{ and } [u, v] \neq [i, j]$$

The expectation vector and covariance matrix of $\tilde{C}_\lambda$ can be calculated accordingly.

After insertion of operation $(1, 1)$, the set of unscheduled jobs is

$$\Omega = \{(1, 2), (2, 1), (2, 2), (3, 1), (3, 2)\}$$

Operation $(2, 1)$ can be considered as the next scheduled operation, and the preceding updating procedure repeats until all operations are scheduled.

Finally, the expectation and variance of the makespan obtained by using the mixture model are as follows:

$$\mu = E[C_{\max}] \approx 23.14101 \qquad \sigma^2 = \mathrm{var}[C_{\max}] \approx 7.90369$$

However, if we use normal distributions to approximate processing times and follow the procedure discussed in Chapter 6, the corresponding approximate values are

$$\mu \approx 23.11057 \qquad \sigma^2 \approx 8.13802$$

To provide a point of reference, we also used simulation to estimate the expectation and variance of the makespan. After 100,000 replications, the sample statistics are

$$\mu = 23.13583 \qquad \sigma^2 = 7.81154$$

Note that the mixture model provides better approximations of the expectation and variance for the makespan of the given schedule than that obtained by using a normal distribution.

8.2.5 Flow-Shop and Parallel-Machine Problems

Since flow shops and parallel machines are special cases of a job-shop machine configuration, schedules for these environments also can be analyzed using mixture models. We expound on this in this section.

8.2.5.1 Flow Shops

A flow shop is a special case of a job shop in which the operations of each job are processed in the same order over the machines. Given the sequence in which the jobs are processed on each machine, the set O_λ can be defined accordingly. Then the MixtureJobShop algorithm can be used to estimate the expectation and variance of the makespan.

Consider a flow-shop makespan problem with three jobs and two machines. Assume the processing time distributions to be the same as those specified in Table 8.10. All the jobs follow the route 1–2 over the machines, and the processing sequence of jobs is assumed to be 1–2–3 on each machine. This is equivalent to a job-shop problem with the routings and processing sequence shown in Tables 8.13 and 8.14.

We approximate processing time of each operation with a mixture of five normal components. By using the mixture model, we obtain expectation and variance of the makespan as follows:

$$\mu = E[C_{\max}] \approx 26.81351 \qquad \sigma^2 = \text{var}[C_{\max}] \approx 10.53852$$

If we use normal distributions to approximate processing times and follow the procedure discussed in Chapter 6, the corresponding approximate values are

$$\mu \approx 26.81997 \qquad \sigma^2 \approx 10.39352$$

Table 8.13. *Routings of Jobs*

Jobs	Routing of a Job on Machines	
Job 1	1	2
Job 2	1	2
Job 3	1	2

Table 8.14. *Machine Processing Sequence*

Machines	Sequence of Jobs on a Machine		
Machine 1	1	2	3
Machine 2	1	2	3

We also used simulation to estimate the expectation and variance of the makespan. After 300,000 replications, the sample statistics are

$$\mu = 26.80963 \qquad \sigma^2 = 10.52031$$

Again, the mixture model provides better approximations of the expectation and variance for the makespan of the given schedule than that obtained using a normal distribution.

8.2.5.2 Parallel Machines Without Preemption

The parallel-machine makespan problem also can be considered as a special case of the job-shop makespan problem with one operation per job. Therefore, the MixtureJobShop algorithm can be applied to estimate the expectation and variance of the makespan for a given schedule on parallel machines as well. Note that in this particular case we have $O_{\lambda J} = \phi$ and $O_\lambda = O_{\lambda M}$. Furthermore, since there is no correlation between completion times of the latest operations on any two different machines, mixture approximations of their distributions can be determined independently. This property can be used to reduce the computational effort needed to approximate the distribution of C_λ.

Consider a schedule of six operations on two parallel machines. The processing time distribution for each operation (designated as a job) is provided in Table 8.15.

In addition, we assume the following assignments of jobs to the machines and the sequences in which they are processed on the machines:

Machine 1:   1–3–4–6
Machine 2:   2–5

We use five-component mixture models to approximate the uniformly distributed processing times. Note that on applying the MixtureJobShop algorithm, the makespan on these two machines can be determined independently. For each machine, the makespan is determined by the completion time of the last job. Hence we can apply the procedure introduced in Section 8.2.3 for single-machine problems. Note that since there are four jobs on machine 1, the mixture distribution of the makespan on machine 1 can be represented as a mixture model with $5 \times 5 \times 5 \times 5 = 625$ normal components. Similarly, the makespan on machine 2 is represented as a mixture model with $5 \times 5 = 25$ components. Therefore, the overall makespan can be calculated by considering

Table 8.15. *Processing Times Distribution*

Job 1	Job 2	Job 3	Job 4	Job 5	Job 6
$U[6,7]$	$U[3,6]$	$U[2,7]$	$U[5,11]$	$U[9,17]$	$U[1,8]$

the 625×25 combinations of machine-wise makespan components. For each of these 625×25 combinations, the overall makespan is the maximum of the makespans on both machines, which can be determined by Clark's formulas. Using this procedure, the mixture models give the following approximation of expectation and variance:

$$\mu = E[C_{max}] \approx 23.59731 \qquad \sigma^2 = var[C_{max}] \approx 8.45454$$

If we use normal distributions to approximate processing times and follow the procedure discussed in Chapter 6, the corresponding approximate values are

$$\mu \approx 23.60658 \qquad \sigma^2 \approx 8.40053$$

We also used Monte Carlo simulation to estimate the expectation and variance of the makespan. After 100,000 replications, the sample statistics are

$$\mu = 23.59672 \qquad \sigma^2 = 8.46845$$

The use of mixture models results in better approximations for the expectation and variance of the makespan than those obtained using normal distributions.

The total completion time problem for parallel machines was already addressed in Section 7.2.2. Note that no assumption has been made there about the processing time distributions.

8.2.5.3 Parallel Machine with Preemptions

As mentioned in Section 7.3, this case can be converted into a job-shop configuration by treating preempted jobs as jobs with multiple operations. Therefore, the MixtureJobShop algorithm still applies. Consider the example depicted in Figure 7.1. We can transfer it into a job-shop schedule with nine jobs and three machines as described in Tables 8.16 and 8.17. The MixtureJobShop algorithm can be applied subsequently.

Table 8.16. *Routings of Jobs*

Jobs	Routing of a Job on Machines		
Job 1	a	b	c
Job 2	a		
Job 3	b	a	b
Job 4	c		
Job 5	b		
Job 6	c		
Job 7	a		
Job 8	b	c	
Job 9	c		

Table 8.17. *Machine Processing Sequence*

Machines	Sequence of Jobs on a Machine				
Machine a	1	2	3	7	
Machine b	3	1	8	5	3
Machine c	6	4	1	8	9

8.2.6 Mixture Reduction

Even though the mixture models give accurate estimates of the expectation and variance of a performance measure for a given schedule, the number of components in such a model increases rather quickly with an increment in the number of operations. This aspect of mixture models leads to high computational complexity and makes them less practical for large-size problems. To circumvent this difficulty, we consider the method of mixture reduction. As the name suggests, mixture reduction seeks to reduce the number of components in a mixture model while keeping the new distribution as close as possible to the original one. Our aim is to embed the method of mixture reduction in our approximation procedure so that we can keep the number of components at an acceptable level without significantly affecting the quality of approximation. In the following discussion we assume that the original mixture model is given by

$$f(\mathbf{x}; \Psi_h) = \sum_{i=1}^{g_h} p_i \varphi(\mathbf{x}; \boldsymbol{\mu}_i, \mathbf{S}_i)$$

where $\varphi(\cdot; \boldsymbol{\mu}_i, \mathbf{S}_i)$ represents the p.d.f. of multivariate normal with expectation vector $\boldsymbol{\mu}_i$ and covariance matrix $\mathbf{S}_i$.

8.2.6.1 Salmond's Approach
Consider the covariance matrix of the original mixture

$$\mathbf{S} = \text{cov}(\mathbf{X}) = \mathbf{W} + \mathbf{B} = \sum_{i=1}^{g_h} p_i \mathbf{S}_i + \sum_{i=1}^{g_h} p_i (\boldsymbol{\mu}_i - \boldsymbol{\mu})(\boldsymbol{\mu}_i - \boldsymbol{\mu})^T$$

where $\boldsymbol{\mu} = \sum_{i=1}^{g_h} p_i \boldsymbol{\mu}_i$ is the expectation vector of the overall distribution. Note that the covariance matrix $\mathbf{S}$ is the sum of two parts. The first part ($\mathbf{W}$) belongs to the covariance within each mixture component, whereas the second part ($\mathbf{B}$) can be regarded as the covariance between mixture components. Now suppose that the ith and jth components are merged into one component, which we denote as $\mathbf{X}_{i,j}$. To preserve the expectation and covariance of the mixture, the weight, expectation, and covariance of this new component are given by

$$p_{i,j} = p_i + p_j$$

$$\mu_{i,j} = E[\mathbf{X}_{i,j}] = \frac{1}{p_{i,j}}(p_i\mu_i + p_j\mu_j)$$

$$\mathbf{S}_{i,j} = \text{cov}[\mathbf{X}_{i,j}]$$

$$= \frac{1}{p_{i,j}}(p_i\mathbf{S}_i + p_j\mathbf{S}_j) + \frac{p_i}{p_{i,j}} \cdot \frac{p_j}{p_{i,j}} \cdot (\mu_i - \mu)(\mu_j - \mu)^T \qquad (8.40)$$

Note that the second term in the expression for $\mathbf{S}_{i,j}$ will be multiplied by $p_{i,j}$ in the calculation of $\mathbf{W}$. The product serves to compensate for the decrement in $\mathbf{B}$ during the merger of components i and j. The value of this decrement is determined as follows: Since

$$\mu_{i,j} - \mu = \frac{1}{p_{i,j}}(p_i\mu_i + p_j\mu_j) - \mu$$

$$= \frac{1}{p_{i,j}}[p_i(\mu_i - \mu) + p_j(\mu_j - \mu)]$$

we have

$$p_{i,j}(\mu_{i,j} - \mu)(\mu_{i,j} - \mu)^T$$

$$= \frac{1}{p_{i,j}}[p_i(\mu_i - \mu) + p_j(\mu_j - \mu)][p_i(\mu_i - \mu) + p_j(\mu_j - \mu)]^T$$

$$= \frac{1}{p_{i,j}}[p_i^2(\mu_i - \mu)(\mu_i - \mu)^T + p_ip_j(\mu_i - \mu)(\mu_j - \mu)^T$$

$$+ p_ip_j(\mu_j - \mu)(\mu_i - \mu)^T + p_j^2(\mu_j - \mu)(\mu_j - \mu)^T]$$

Therefore, the decrement in $\mathbf{B}$ is given by

$$\Delta B = p_i(\mu_i - \mu)(\mu_i - \mu)^T + p_j(\mu_j - \mu)(\mu_j - \mu)^T - p_{i,j}(\mu_{i,j} - \mu)(\mu_{i,j} - \mu)^T$$

$$= \frac{1}{p_{i,j}}[p_{i,j}p_i(\mu_i - \mu)(\mu_i - \mu)^T + p_{i,j}p_j(\mu_j - \mu)(\mu_j - \mu)^T$$

$$- p_i^2(\mu_i - \mu)(\mu_i - \mu)^T - p_ip_j(\mu_i - \mu)(\mu_j - \mu)^T$$

$$- p_ip_j(\mu_j - \mu)(\mu_i - \mu)^T - p_j^2(\mu_j - \mu)(\mu_j - \mu)^T]$$

$$= \frac{1}{p_{i,j}}[p_jp_i(\mu_i - \mu)(\mu_i - \mu)^T + p_ip_j(\mu_j - \mu)(\mu_j - \mu)^T$$

$$- p_ip_j(\mu_i - \mu)(\mu_j - \mu)^T - p_ip_j(\mu_j - \mu)(\mu_i - \mu)^T]$$

$$= \frac{p_ip_j}{p_{i,j}}[(\mu_i - \mu) - (\mu_j - \mu)][(\mu_i - \mu) - (\mu_j - \mu)]^T$$

$$= \frac{p_ip_j}{p_{i,j}}(\mu_i - \mu_j)(\mu_i - \mu_j)^T$$

If the value of ΔB is relatively small, less change will be incurred in the partition of within-component and between-component covariances (i.e., $\mathbf{W}$ and $\mathbf{B}$). On the basis of this observation, Salmond (1988) used the following cost function to select the merging components so that less change in $\mathbf{B}$ (or, equivalently, $\mathbf{W}$) is incurred:

$$d_S^2 = tr(\mathbf{S}^{-1}\Delta B) = \frac{p_i p_j}{p_{i,j}}(\boldsymbol{\mu}_i - \boldsymbol{\mu}_j)^T \mathbf{S}^{-1}(\boldsymbol{\mu}_i - \boldsymbol{\mu}_j)$$

Note that ΔB has been multiplied by $\mathbf{S}^{-1}$ so that the cost function d_S^2 is invariant under linear transformations of vector $\mathbf{X}$. On the basis of this definition, Salmond's joining algorithm of mixture reduction can be described as follows: A pair of components is selected to merge at every iteration of the algorithm so that the cost function is minimized myopically. The algorithm stops if the number of components achieves a prescribed value or if the next merger will result in an unacceptably high value of d_S^2.

Since the joining algorithm is a greedy algorithm, it does not guarantee that the final mixture achieves the minimal change in $\mathbf{B}$ among all the possible mixture models with the same number of components. Another drawback of this approach is that the cost function d_S^2 does not depend on the individual covariance matrices of the two merging components. Hence two components may be selected to merge even if they are very different in shape.

8.2.6.2 Williams' Approach

The reduction of a mixture model can be viewed as an optimization problem. The objective is to minimize a distance measure between the reduced distribution and the original distribution. The distance measure should be straightforward to evaluate. It is also desirable that the derivatives of this distance measure can be expressed in closed form so that the optimization problem is easy to solve. To that end, we briefly introduce the approach of Gaussian mixture reduction proposed by Williams (2003).

The distance measure adopted is called the *integral square difference* (ISD) *measure* and is defined as

$$d_W = \int (f(x, \Psi_h) - f(x, \Psi_r))^2 dx$$

where $f(\cdot)$ represents the p.d.f. of mixture normal; vectors Ψ_h and Ψ_r are, respectively, the parameters (weights, mean, and covariance) of the mixture before and after the reduction. Next, we will show that both d_W and its derivatives can be written in closed form. Hence we can use gradient-descent methods of nonlinear optimization to determine the reduced mixture distribution iteratively.

Note that the ISD measure can be written in its expanded form as follows:

$$dW = d_{hh} + d_{rr} - 2d_{hr}$$

where

$$d_{hh} = \int (f(\mathbf{x}; \Psi_h))^2 \, d\mathbf{x}$$

$$d_{rr} = \int (f(\mathbf{x}; \Psi_r))^2 \, d\mathbf{x}$$

$$d_{hr} = \int f(\mathbf{x}; \Psi_h) f(\mathbf{x}; \Psi_r) \, d\mathbf{x}$$

The expanded form of mixture densities includes

$$f(\mathbf{x}; \Psi_h) = \sum_{i=1}^{g_h} p_i \varphi(\mathbf{x}; \boldsymbol{\mu}_i, \mathbf{S}_i)$$

$$f(\mathbf{x}; \Psi_r) = \sum_{i=1}^{g_r} \bar{p}_i \varphi(\mathbf{x}; \bar{\boldsymbol{\mu}}_i, \bar{\mathbf{S}}_i)$$

where $\varphi(x; \cdot, \cdot)$ represents the p.d.f. of multivariate normal.

Willams (2003) has further proved that

$$d_{hh} = \sum_{i=1}^{g_h} \sum_{j=1}^{g_h} p_i p_j \varphi(\boldsymbol{\mu}_i; \boldsymbol{\mu}_j, \mathbf{S}_i + \mathbf{S}_j)$$

$$d_{rr} = \sum_{i=1}^{g_r} \sum_{j=1}^{g_r} \bar{p}_i \bar{p}_j \varphi(\bar{\boldsymbol{\mu}}_i; \bar{\boldsymbol{\mu}}_j, \bar{\mathbf{S}}_i + \bar{\mathbf{S}}_j)$$

$$d_{hr} = \sum_{i=1}^{g_h} \sum_{j=1}^{g_r} p_i \bar{p}_j \varphi(\boldsymbol{\mu}_i; \bar{\boldsymbol{\mu}}_j, \mathbf{S}_i + \bar{\mathbf{S}}_j)$$

To derive derivatives of the preceding distance terms, two substitutions are made, namely,

$$\bar{p}_i = \frac{q_i^2}{\sum_{i=1}^{g_r} q_i^2} \quad \text{and} \quad \bar{\mathbf{S}}_i = \mathbf{L}_i \mathbf{L}_i^T$$

This leads to the following results:

$$\frac{\partial d_{rr}}{\partial q_j} = 4q_j \sum_{i=1}^{g_r} q_i^2 \varphi(\bar{\boldsymbol{\mu}}_i; \bar{\boldsymbol{\mu}}_j, \mathbf{L}_i \mathbf{L}_i^T + \mathbf{L}_j \mathbf{L}_j^T)$$

$$\frac{\partial d_{hr}}{\partial q_j} = 2q_j \sum_{i=1}^{g_h} p_i \varphi(\boldsymbol{\mu}_i; \bar{\boldsymbol{\mu}}_j, \mathbf{S}_i + \mathbf{L}_j \mathbf{L}_j^T)$$

$$\frac{\partial d_{rr}}{\partial \bar{\mu}_j} = -2q_j^2 \sum_{i=1}^{g_r} q_i^2 (\mathbf{L}_i \mathbf{L}_i^T + \mathbf{L}_j \mathbf{L}_j^T)^{-1} (\bar{\mu}_j - \bar{\mu}_i) \varphi(\bar{\mu}_i; \bar{\mu}_j, \mathbf{L}_i \mathbf{L}_i^T + \mathbf{L}_j \mathbf{L}_j^T)$$

$$\frac{\partial d_{hr}}{\partial \bar{\mu}_j} = -2q_j^2 \sum_{i=1}^{g_h} p_i (\mathbf{S}_i + \mathbf{L}_j \mathbf{L}_j^T)^{-1} (\bar{\mu}_j - \mu_i) \varphi(\mu_i; \bar{\mu}_j, \mathbf{S}_i + \mathbf{L}_j \mathbf{L}_j^T)$$

$$\frac{\partial d_{rr}}{\partial \mathbf{L}_j} = 2q_j^2 \sum_{i=1}^{g_r} q_i^2 \varphi(\bar{\mu}_i; \bar{\mu}_j, \mathbf{L}_i \mathbf{L}_i^T + \mathbf{L}_j \mathbf{L}_j^T)(\mathbf{L}_i \mathbf{L}_i^T + \mathbf{L}_j \mathbf{L}_j^T)^{-1}$$

$$\cdot [(\bar{\mu}_j - \bar{\mu}_i)(\bar{\mu}_j - \bar{\mu}_i)^T - (\mathbf{L}_i \mathbf{L}_i^T + \mathbf{L}_j \mathbf{L}_j^T)] \cdot (\mathbf{L}_i \mathbf{L}_i^T + \mathbf{L}_j \mathbf{L}_j^T)^{-1} \mathbf{L}_j$$

$$\frac{\partial d_{hr}}{\partial \mathbf{L}_j} = q_j^2 \sum_{i=1}^{g_h} p_i \varphi(\mu_i; \bar{\mu}_j, \mathbf{S}_i + \mathbf{L}_j \mathbf{L}_j^T)(\mathbf{P}_i + \mathbf{L}_j \mathbf{L}_j^T)^{-1}$$

$$\cdot [(\bar{\mu}_j - \mu_i)(\bar{\mu}_j - \mu_i)^T - (\mathbf{P}_i + \mathbf{L}_j \mathbf{L}_j^T)] \cdot (\mathbf{P}_i + \mathbf{L}_j \mathbf{L}_j^T)^{-1} \mathbf{L}_j \qquad (8.41)$$

Given the preceding expressions, the nonlinear optimization problem can be solved iteratively in the space of $\{q_i, \bar{\mu}_i, \mathbf{L}_i\}$, $i = 1, \ldots, g_r$. The initial solution is obtained by using a greedy procedure that is similar to Salmond's joining algorithm but with d_W replacing the cost function d_S^2. For further details of the optimization process, interested readers are referred to Williams (2003).

For our objective of estimating the expectation and variance of a performance measure for a given schedule, it is also preferable to maintain the first two moments of the mixture after the reduction. To achieve this, we can add the following constraints to the optimization problem:

$$\sum_{i=1}^{g_r} \bar{p}_i \bar{\mu}_i = \mu_o \qquad (8.42)$$

$$\sum_{i=1}^{g_r} \bar{p}_i \bar{\mathbf{S}}_i + \sum_{i=1}^{g_r} \bar{p}_i (\bar{\mu}_i - \mu_o)(\bar{\mu}_i - \mu_o)^T = \mathbf{S}_o \qquad (8.43)$$

where $\mu_o = \sum_{i=1}^{g_h} p_i \mu_i$ and $\mathbf{S}_o = \sum_{i=1}^{g_h} p_i \mathbf{S}_i + \sum_{i=1}^{g_h} p_i (\mu_i - \mu_o)(\mu_i - \mu_o)^T$, respectively, are the expectation vector and covariance matrix of the original mixture.

Although the approach based on ISD is of theoretical interest, it is not attractive from computational point of view. First of all, note that the derivatives in Equation (8.41) involve matrix inversion and evaluation of a normal density function, both of which are computationally demanding for high-dimensional data. Second, if we add constraints (8.42) and (8.43) to preserve the first two moments of the mixture, the problem becomes a constrained global optimization problem, which is harder to solve. Therefore, in the numerical example provided at the end of this section, we do not include these extra constraints.

8.2.6.3 Runnalls' Approach

Runnalls' (2007) algorithm for mixture reduction is similar to Salmond's joining algorithm. At each iteration of the algorithm, two components are selected to merge in accordance with a cost function. However, the cost function that Runnalls adopted is based on the Kullback–Leibler (KL) discrimination, which is defined as

$$d_{KL}(f_1(\mathbf{x}), f_2(\mathbf{x})) = \int f_1(\mathbf{x}) \log \frac{f_1(\mathbf{x})}{f_2(\mathbf{x})} d\mathbf{x}$$

This is also called the *Kullback–Leibler divergence* of $f_2(\mathbf{x})$ from $f_1(\mathbf{x})$. In information theory, it is used to represent the difference between an approximation and its theoretical model. Despite its intuitive interpretation, the KL measure is hard to apply in our problem. The difficulty is that when $f_1(\mathbf{x})$ and $f_2(\mathbf{x})$ are both mixtures of normal distributions, d_{KL} cannot be expressed in closed form. To circumvent this problem, Runnalls proposed the following dissimilarity measure to represent the divergence of the reduced mixture from the original mixture after the ith and jth components are merged:

$$B_{i,j} = \tfrac{1}{2}[p_{i,j} \log \det(\mathbf{S}_{i,j}) - p_i \log \det(\mathbf{S}_i) - p_j \log \det(\mathbf{S}_j)]$$

Note that $p_{i,j}$ and $\mathbf{S}_{i,j}$ are, respectively, the weight and covariance matrix of the new component derived from components i and j [see Equation (8.40)].

Runnalls (2007) has shown that the dissimilarity measure $B_{i,j}$ is an upper bound for the Kullback–Leibler divergence. That is,

$$B_{i,j} \geq d_{kl}(f(\mathbf{x}; \Psi_h), f(\mathbf{x}; \Psi_r))$$

where $f(\mathbf{x}; \Psi_h)$ and $f(\mathbf{x}; \Psi_r)$ are, respectively, the p.d.f. before and after the merger of the ith and jth mixing components. Note that, by definition, $B_{i,j}$ is a symmetric measure (i.e., $B_{i,j} = B_{j,i}$). Its value is related to both the expectation and the variance of the two merging components. Furthermore, $B_{i,j} = 0$ if and only if at least one of the following statements is true: (1) $p_i = 0$, (2) $p_i = 0$, or (3) $\mu_i = \mu_j$ and $\mathbf{S}_i = \mathbf{S}_j$. Hence $B_{i,j}$ is a viable surrogate for the Kullback–Leibler divergence. Using $B_{i,j}$ as the cost function instead of d_S^2, we can reduce the number of components iteratively in a way that is similar to Salmond's joining algorithm.

To evaluate the effectiveness of the aforementioned reduction approaches, we reconsider the job-shop problem presented in Tables 8.10 through 8.12. The mixture-reduction procedure is incorporated into the MixtureJobShop algorithm by changing step 2 as follows:

Step 2. Let $\tilde{O}_\lambda = (O_\lambda - \{O_{[p,q]}, O_{[i,j-1]}\}) \cup \{O_{[i,j]}\}$

Table 8.18. *Results for Job-Shop Makespan Problem with Mixture Reduction*

Mixture Reduction	$E[C_{max}]$	$var[C_{max}]$	CPU Time* (in seconds)
No reduction	23.14101	7.90369	21.89
Salmond's approach	23.13773	7.96249	0.28
Williams' approach	23.15478	8.02265	374.30
Runnalls' approach	23.14288	7.99813	0.31
Normal distribution	23.11057	8.13802	*negligible*
Simulation results	23.14544	7.89138	97.58

* CPU times are obtained using Matlab 7.6 on a 2.66-GHz Intel Core2 Duo Computer with 2 GB of memory.

Determine the mixture distribution of $\tilde{\mathbf{C}}_\lambda$ according to the distributions of $\mathbf{C}_\lambda$ and $P_{[i,j]}$.

If the number of mixing components in $\tilde{\mathbf{C}}_\lambda$ exceeds the value of G_0, apply a mixture-reduction algorithm on $\tilde{\mathbf{C}}_\lambda$ to reduce the number of components to G_0.

Again, we approximate the processing time of each operation with a mixture of five normal components. The parameter associated with the mixture-reduction algorithm is set at $G_0 = 10$. We summarize the results in Table 8.18. Previous results for the same problem from Section 8.2.4 are also presented in the table for comparison.

Note that by using each of the three reduction approaches, we are able to obtain better approximations of the expectation and variance than that for a normal distribution. There is no significant difference between the results obtained by the three reduction approaches. However, Williams' approach requires a much longer CPU time than that required by the other two. Its CPU time is even longer than the time needed when the original mixtures are used (without any reduction). On the other hand, by incorporating Salmond's or Runnalls' approach of mixture reduction into the MixtureJobShop algorithm, the CPU time is reduced substantially without much effect on the accuracy of the results.

8.2.7 Application to Stochastic Activity Network

An activity network (Elmaghraby, 1977) (or a project network) is a network graph that depicts the precedence relationships among the activities of a project. There are two forms of activity networks: activity-on-node and activity-on-arc. As an example, consider the following project evaluation and review technique (PERT) network (Figure 8.10). It shows an activity-on-arc network. Nodes represent the events (start and end of activities) that are incurred over the duration of the project. They are labeled with numbers in 10s to allow insertion of addition events.

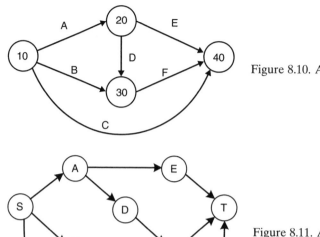

Figure 8.10. A PERT network.

Figure 8.11. An activity-on-node network.

For this particular example, nodes 10 and 40 represent the start and end of the project, respectively. Arcs are labeled as $A - F$ to denote the activities, each of which requires some amount of time to finish. Consider node 30, for example. The incoming arcs B and D are prerequisite activities for this event; that is event 30 will be realized only after both these activities are finished. On the other hand, its outgoing arc/activity F will be started once event 30 occurs.

The PERT network in Figure 8.10 can be equivalently converted into an activity-on-node network as depicted in Figure 8.11. Note that nodes in this new network are labeled with activity names that correspond to the arcs in Figure 8.10. The starting and ending nodes are labeled as S and T, respectively. Their durations are assumed to be zero. Note that in an activity-on-node network, the arcs represent the precedence relationships among the activities (nodes).

As a modeling tool, activity networks are used commonly for time management of complex projects. Their application also extends to timing analysis of integrated circuits. To establish the correspondence between an electronic circuit and an activity network, signal delays caused by gates and interconnecting cables are treated as durations of activities, whereas the structure of the activity network is dictated by the circuit design.

In its basic form, an activity network is assumed to have deterministic activity durations. For example, although PERT assumes that activity durations follow beta distributions, it uses expected values of activity durations to determine the critical path of the project. The calculation of earliest start and latest finish times for each activity is also based on expected values of durations. In other words, the stochasticity is suppressed by taking the expectation of activity duration; the resulting network is treated as a deterministic problem. Note that PERT uses the following approximation to evaluate the expected duration of a given activity:

$$T_E = \frac{O + 4M + P}{6}$$

where O is the optimistic (minimum) time, P is the pessimistic (maximum) time, and M is the most likely time. Similarly, in static timing analysis of integrated circuits, all delays are assumed to be deterministic.

Deterministic activity networks are relatively easy to analyze. The makespan of a given project (or the maximum circuit delay of an integrated circuit) can be found by solving the longest-path problem. However, for real-life problems, activity durations are unlikely to be deterministic. In a project, the completion of a given task is affected by multiple factors, such as the availability of resources, shared personnel, weather, etc. For an integrated circuit, the same design may lead to different electrical performances owing to the inevitable randomness in the wafer fabrication process. Using deterministic activity durations to determine the longest path will not yield accurate results for the expected value of the makespan, let alone capture its intrinsic stochasticity. In fact, it is well known that the makespan calculated by using the expected values of durations always underestimates the expected value of the actual makespan (a random variable).

Stochastic activity network is a direct extension of its deterministic counterpart. Instead of deterministic values, the activity durations are described by probability distributions. To evaluate the stochastic makespan, several approaches are available [see Yao and Chu (2007) and references cited therein]. These are

1. Exact analysis
2. Approximation
3. Monte Carlo simulation

Exact analysis, when tractable, can provide the most accurate results. However, if activity durations follow general distributions, it is very difficult to obtain analytical results. The difficulty arises for two reasons. First, since the evaluation of the length of a path (from the starting node to the ending node) involves a convolution of multiple durations, a multivariate integral is needed for its determination. This integral is rarely reducible to any closed form, except for some special distributions of activity durations. Second, the desired makespan is determined by the maximum of all possible paths from the starting node to the ending node. This maximum is difficult to evaluate owing to a large number of possible paths and the correlations that occur between them. Note that even if activity durations are distributed independently, the lengths of two paths will be correlated random variables if they share a common set of activities. The Monte Carlo simulation method is readily applicable to evaluation of the stochastic makespan. However, it generally requires a large number of simulation runs to yield results with acceptable confidence levels. Therefore, we focus on the second approach and strive to provide good approximations of expectation and variance for the makespan with moderate computational effort.

Two types of approximation methods have been presented in the literature. The first type of these methods relies on discrete representation of activity durations (see Shogan, 1977; Dodin, 1985; Yao and Chu, 2007). For activity durations that are continuous random variables, discretization schemes are applied to convert them into discrete variables (also see Agrawal and Elmaghraby, 2001). Discrete representation permits numerical evaluation of the summation and maximum of random variables. Various schemes have been devised to reduce the computational overhead and the error introduced owing to correlations. The second type of method promotes the use of a normal distribution to represent activity completion times (see Sculli, 1983). Clark's formulas can be used to determine the start time of any given activity from the completion time distributions of its preceding activities (as we illustrated in Chapter 6 for the case of job-shop problems). However, for methods presented in the context of activity networks, possible correlations among different paths are simply ignored. Soroush (1994) assumed the length of a given path to be normally distributed (by the central limit theorem). The probability that the project duration exceeds some prescribed time limit is approximately calculated by solving a deterministic optimal-path problem with fractional objective function. Wang and Mazumder (2005) used a multivariate normal to model the joint distribution of intermediate random variables in the network. They also derived formulas to determine expectation and variance of the maximum for a multivariate normal distribution.

In what follows, we present an alternative approach to approximate the distribution of the makespan for a given project using mixture normal representation of activity durations. Owing to its compact support and controllable shape, the beta distribution has been used commonly in PERT analysis to represent activity durations. Hence, as a preparation for our approximation procedure, we demonstrate the use of the EM algorithm to fit normal mixtures to beta distributions for various parameter values. This is illustrated in Figure 8.12, where the mixture models obtained for four beta distributions are depicted. The corresponding beta distributions are drawn in the background with dashed lines. Note that the mixture models accurately represent beta distributions of different parameter values with increments in the number of components.

8.2.7.1 Approximation Procedure

In the first step, we convert the activity network of interest to an equivalent form to enable use of an algorithm similar to MixtureJobShop to derive approximate values of expectation and variance for the project makespan. The goal of this conversion is to have at most two incoming arcs for each node. Consider an activity-on-arc network. We first find nodes with more than two incoming arcs. For each of these nodes, the incoming arcs are reorganized in a binary tree structure with possible additions of nodes and arcs to convert it to the

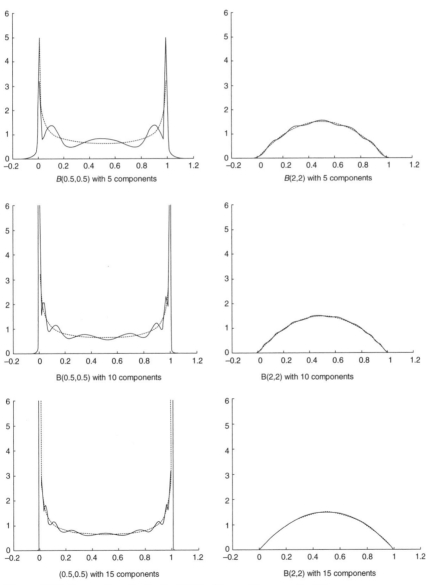

Figure 8.12. Approximating beta distributions with normal mixtures.

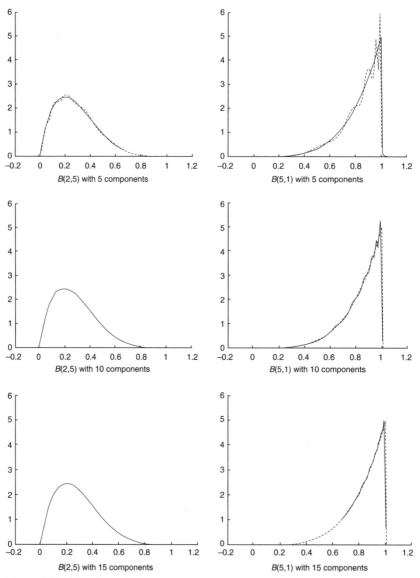

Figure 8.12. (Continued)

desired form. For instance, consider the network in Figure 8.10. Node 40 is the only node with more than two incoming arcs. The network after conversion is depicted in Figure 8.13. Note that a new node, 35, is inserted to represent completion time of activities E and F. The additional arc connecting 35 and 40 is labeled P. It is a pseudoactivity with zero duration.

In case of an activity-on-node network, a similar procedure can be applied to reduce the maximum number of incoming arcs at a node to two. This is depicted below (see Figure 8.14) for the network in Figure 8.11. Again, the added activity P has zero duration.

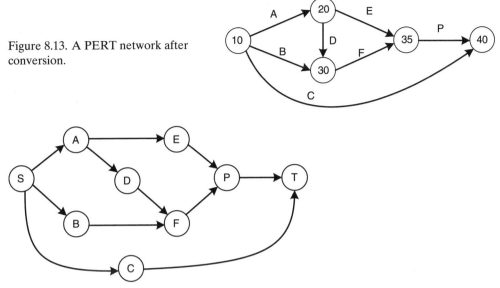

Figure 8.13. A PERT network after conversion.

Figure 8.14. An activity-on-node after conversion.

Next, we present our approximation algorithm for an activity-on-node network. The approximation algorithm for an activity-on-arc network will be discussed later with minor modifications. The basic idea of this approximation algorithm is similar to that of the MixtureJobShop algorithm with operations replaced by nodes/activities. Starting from node S, activities in the network are scheduled one node at a time. A node is regarded as schedulable only if all its preceding nodes have been scheduled. The completion time of a newly scheduled node is determined by using the joint completion time distribution of previously scheduled nodes and the distribution of its own duration. Both these distributions are represented as mixture normal distributions. The completion times of previously scheduled nodes have a multivariate distribution. Not all these random variables are directly related to completion of the newly scheduled node. To reduce storage requirements and computational overhead, we define the set of "latest nodes" N_λ in a way similar to the definition of "latest operations" for a job-shop environment as follows:

$$N_\lambda = \{a | a \in N_S; \exists (a, b) \in A, b \notin N_S\}$$

where N_S is the set of previously scheduled nodes, and A is the set of directed arcs. Referring to Figure 8.14, suppose that nodes $S, A, B, C,$ and D have been scheduled. The set of "latest nodes" is $N_\lambda = \{A, B, C, D\}$. In this case, the completion time of node F can be determined as

$$C_F = \max\{C_B, C_D\} + P_F \tag{8.44}$$

where C and P represent completion time and duration of a given activity, respectively.

Since Equation (8.44) closely resembles Equation (8.39) for a job shop, the updating scheme for the joint distribution of $\mathbf{C}_\lambda$ (see Section 8.2.4) can be applied here for the completion times of the nodes in N_λ. The only difference is in the way the set N_λ is updated after the scheduling of a new node. Note that the completion time of the last node of the network is the makespan of the project. The expectation and variance of the makespan can be determined from their component-wise values by using the law of total expectation and the law of total variation.

For an activity-on-arc network, we need to make the following modifications to the preceding approximation algorithm. Note that the set of "latest nodes" now corresponds to the latest events in the network. The time of occurrence of a newly scheduled event is the maximum among the completion times of the activities on its incoming arcs. For instance, consider node 30 in Figure 8.13. The occurrence time of this event is given by

$$C_{30} = \max\{C_{10} + P_B, C_{20} + P_D\} \tag{8.45}$$

We can assume the particular mixture components selected for P_B and P_D and the joint distribution of (C_{10}, C_{20}) to follow normal distributions. Note that the expectation vector and covariance matrix of (C_{10}, C_{20}) are available from previous steps of the approximation algorithm. The expectation and variance of P_B and P_D are assumed to be given as input. Furthermore, we have

$$E[C_{10} + P_B] = E[C_{10}] + E[P_B]$$

$$E[C_{20} + P_D] = E[C_{20}] + E[P_D]$$

$$\mathrm{cov}[C_{10} + P_B, C_{20} + P_D]$$

$$= \mathrm{cov}[C_{10}, C_{20}] + \mathrm{cov}[P_B, C_{20}] + \mathrm{cov}[C_{10}, P_D] + \mathrm{cov}[P_B, P_D]$$

$$= \mathrm{cov}[C_{10}, C_{20}] + 0 + 0 + 0 \quad \text{(owing to independence of activity durations)}$$

$$= \mathrm{cov}[C_{10}, C_{20}]$$

Using the preceding three terms, Clark's formulas can be applied to derive the expectation and variance of C_{30}. The covariance between C_{30} and the occurrence time C_n of another event n in the set N_λ also can be derived by Clark's formula, noting that

$$\mathrm{cov}(C_{10} + P_B, C_n)$$
$$= \mathrm{cov}(C_{10}, C_n) + \mathrm{cov}(P_B, C_n)$$
$$= \mathrm{cov}(C_{10}, C_n) + 0$$
$$= \mathrm{cov}(C_{10}, C_n)$$

and

$$\mathrm{cov}(C_{20} + P_D, C_n)$$
$$= \mathrm{cov}(C_{20}, C_n) + \mathrm{cov}(P_D, C_n)$$
$$= \mathrm{cov}(C_{20}, C_n) + 0$$
$$= \mathrm{cov}(C_{20}, C_n)$$

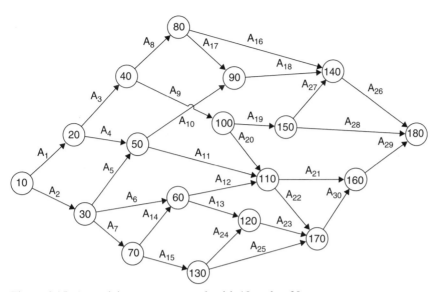

Figure 8.15. An activity-on-arc network with 18 nodes, 30 arcs.

where $\mathrm{cov}(C_{10}, C_n)$ and $\mathrm{cov}(C_{20}, C_n)$ are determined in previous steps of the algorithm.

We repeat the preceding procedure for every combination of mixture components from the distributions of (C_{10}, C_{20}), P_B, and P_D to get the mixture representation of the completion times in the updated set of N_λ.

To illustrate the preceding procedure, we approximate the expectation and variance of the makespan for the activity-on-arc network depicted in Figure 8.15. The activity durations are

$$A_i = d_i + r_i \cdot B_i$$

where B_i assumes a beta distribution with parameters α_i and β_i. Values of these parameters are randomly generated as shown in Table 8.19. We approximate the duration of each activity with a mixture of five normal components. To reduce computational effort, we use Runnall's method of mixture reduction on the mixture model of C_λ (the completion times of the nodes in N_λ) so that the number of mixture components does not exceed 10. The expectation and variance of the makespan obtained by using the mixture model are as follows:

$$\mu = E[\text{Makespan}] \approx 53.62871 \qquad \sigma^2 = \mathrm{var}[\text{Makespan}] \approx 3.08303$$

However, if we use normal distributions to approximate activity durations and follow the same procedure (normal distribution can be regarded as a mixture model with only one component), the corresponding approximate values are

$$\mu \approx 53.63340 \qquad \sigma^2 \approx 3.17731$$

Table 8.19. *Parameters of Activity Durations*

Activity(i)	d_i	r_i	α_i	β_i
1	6.5	5	2	5
2	3.5	3	5	1
3	2.5	5	2	5
4	3	4	2	2
5	7.5	3	2	2
6	3.5	5	2	5
7	3.5	3	5	1
8	3.5	3	2	5
9	7.5	3	2	5
10	5.5	5	0.5	0.5
11	5	4	2	5
12	3.5	5	5	1
13	5.5	3	0.5	0.5
14	7.5	3	2	2
15	3	6	0.5	0.5
16	7.5	5	2	5
17	6.5	5	2	5
18	4	4	2	2
19	7	4	5	1
20	6.5	3	2	2
21	5.5	5	2	5
22	2	6	2	5
23	4	6	2	2
24	2	6	0.5	0.5
25	4.5	5	5	1
26	6	6	5	1
27	6	6	5	1
28	3	4	2	2
29	5	4	0.5	0.5
30	7.5	5	2	2

To provide a point of reference, we also used simulation to estimate the expectation and variance of the makespan. After 100,000 replications, the sample statistics are

$$\mu = 53.62810 \qquad \sigma^2 = 3.11604$$

Note that the mixture model provides better approximations of the expectation and variance for the makespan of the given schedule than that obtained by using a normal distribution. Using simulation results as a benchmark, the relative errors of mixture approximation are, respectively, 0.0011% for expectation and 1.1% for variance. The relative errors of normal approximation are, respectively, 0.0099% for expectation and 2.0% for variance.

9 Concluding Remarks

The primary motivation for the work presented in this book has been to address the inherent variability of various parameters (e.g., job processing times and due dates, among others) that are encountered in the scheduling of tasks in a production environment and their significant impact on system performance. Variability in scheduling has been modeled in the form of random variables, and most of the work undertaken in this field has considered optimizing the expected value of the performance measure of interest. In doing so, very little importance was attributed to variability of the system, and consequently, very little significance was attributed to the variance of a performance measure. It is often seen that a scheduler's preference is not only for a schedule that has a lower expected value but also for a schedule that provides minimum variability. Hence, schedule evaluation must be performed in terms of both the expectation and the variance of the performance measure. The inclusion of variance in schedule optimization leads to complexities that, apparently, have resulted in a dearth of reported work in the literature as far as evaluating the variance of a performance measure is concerned. Our aim has been to develop closed-form expressions (wherever possible) and devise methodologies to determine the expectation and variance for different performance measures and in diverse scheduling environments, in the face of processing time variability. This effort would lead the way in considering variance issues in schedule optimization and determining schedules that are both expectation- and variance-efficient.

Along the way, we also have presented other published work that involved simultaneous consideration of expectation and variance in stochastic scheduling. Two approaches stand out in this regard: (1) robust scheduling and (2) generation of nondominated schedules. The robust scheduling approach attempts to hedge against processing time uncertainty by determining "robust schedules" that can withstand the variability inherent in the job processing times and provide optimal performance in the long run. We have presented in detail the various robust scheduling model formulations and solution methodologies that are available in the literature. As the name indicates, the second approach strives to generate a set of possible "candidate schedules." The user then can select the schedule that is applicable to his or her environment. The schedules comprising this set are commonly called *nondominated schedules* or *expectation-variance (EV)–efficient schedules*. There are also a few methodologies available to select a "preferred schedule" from this set based on user-specified requirements. We have provided a detailed documentation of the various models and solution approaches that have been reported in the literature in this regard.

We have then continued with our EV analysis where we considered the following scheduling environments and have analytically evaluated (wherever possible) the expectation and variance of different performance measures:

1. Scheduling on a single machine
2. Scheduling on identical machines in parallel
3. Permutation flow shops with unlimited intermediate storage
4. Job shops with unlimited intermediate storage
5. Activity networks.

The performance measures that we have considered include both completion-time-based and due-date-based measures for the single-machine problem, whereas the makespan objective is analyzed for the flow-shop and job-shop cases. For parallel-machine models, the measures used are total completion time and makespan for the case of no preemption and only makespan for the preemption case. The processing times are stochastic with known probability distributions, and the schedule is assumed to be known *a priori*. Computing the variance of different performance measures is not a straightforward task. For instance, the analysis of tardiness-based performance measures for the single-machine problem involves use of the maximum operator, and mathematical theory does not allow for exact evaluation of the maximum function. Hence Clark's equations (Clark, 1961) for normally distributed random variables are adapted in our analysis to determine approximate evaluations of the performance measures. A similar approach is adopted for makespan-related measures for parallel-machine, flow-shop, and job-shop environments as well. In addition, evaluation of the correlations between the various random variables becomes a challenging but essential part of this analysis. The correlation structures become progressively complex as we shift from a single-machine to a multimachine environment. We have also extended our analysis to the case of general processing times. We have developed a new approach based on the use of mixture models to evaluate the expectation and variance of the performance measures of a given schedule for various machine configurations. For the machine configurations and performance measures considered, this new approach is able to yield more accurate results in case the processing time distributions differ from a normal distribution. It provides a useful methodology for application in real-life environments.

9.1 Significance of This Work

Variance consideration in stochastic scheduling has been a very challenging field of research, and it holds the interest of many researchers. The significance of our work lies in the fact that we have addressed an important issue that has received limited attention in the literature. We have elicited an important point that schedules must be evaluated in terms of both the expectation and the variance. By incorporating variance issues in scheduling, a scheduler will be in a position to make better decisions after having known both the expected value and variability of a given schedule. This, in turn, will increase the stability of the production system, which is the ultimate goal of all production managers.

In addition, by providing a detailed review on the work by other researchers who have attempted to address this issue, we have provided a comprehensive compilation that deals specifically with EV considerations in stochastic scheduling with uncertainty in processing times.

This book is accompanied by a CD that contains a software named *XVA-Sched*. This is a user-friendly software of all the algorithms (for various problems) developed in Chapters 4 through 8. The instructions on how to use the software are provided in Section A.4 of the Appendix. Our methodologies and software can be easily integrated with the existing scheduling packages to identify EV-efficient schedules. Thus our work directly contributes to the determination of schedules that are good in terms of both the expected value and the variability of the performance measures considered.

APPENDIX

A.1 Analysis for a Single-Machine Total Tardiness Problem

Job Number	1	2	3
Processing time	$U(20, 60)$	$U(10, 30)$	$U(70, 95)$
Due date	58	87	175

Given the processing sequence 1–2–3, we can derive completion time distributions analytically as follows:

$$C_{[1]} = P_{[1]} \sim U(20, 60)$$

Hence $f_{C_{[1]}}(x) = \begin{cases} \dfrac{1}{40} & 20 \le x \le 60 \\ 0 & \text{otherwise} \end{cases}$

$$C_{[2]} = P_{[1]} + P_{[2]}$$

Consider the following partition on the region of support:

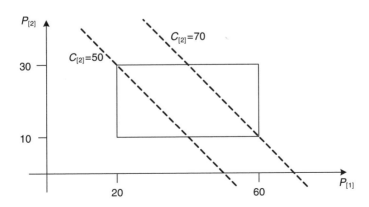

For $x \in [30, 50]$,

$$F_{C_{[2]}}(x) = \Pr\left(C_{[2]} \leq x\right) = \int_{20}^{x-10} \int_{10}^{x-x_1} f_{P_{[2]}}(x_2) dx_2 f_{P_{[1]}}(x_1) dx_1$$

$$= \int_{20}^{x-10} \int_{10}^{x-x_1} \frac{1}{20} dx_2 \frac{1}{40} dx_1$$

$$f_{C_{[2]}}(x) = \frac{x-30}{800}$$

For $x \in [50, 70]$,

$$F_{C_{[2]}}(x) = \Pr\left(C_{[2]} \leq x\right) = \Pr\left(C_{[2]} \leq 50\right) + \Pr\left(50 \leq C_{[2]} \leq x\right)$$

$$= \Pr\left(C_{[2]} \leq 50\right) + \int_{10}^{30} \int_{50-x_2}^{x-x_2} f_{P_{[1]}}(x_1) dx_1 f_{P_{[2]}}(x_2) dx_2$$

$$= \Pr\left(C_{[2]} \leq 50\right) + \int_{10}^{30} \int_{50-x_2}^{x-x_2} \frac{1}{40} dx_1 \frac{1}{20} dx_2$$

$$f_{C_{[2]}}(x) = \frac{1}{40}$$

For $x \in [70, 90]$,

$$F_{C_{[2]}}(x) = \Pr\left(C_{[2]} \leq x\right) = \Pr\left(C_{[2]} \leq 70\right) + \Pr\left(70 \leq C_{[2]} \leq x\right)$$

$$= \Pr\left(C_{[2]} \leq 70\right) + \int_{10}^{x-60} \int_{70-x_2}^{60} f_{P_{[1]}}(x_1) dx_1 f_{P_{[2]}}(x_2) dx_2$$

$$+ \int_{x-60}^{30} \int_{70-x_2}^{x-x_2} f_{P_{[1]}}(x_1) dx_1 f_{P_{[2]}}(x_2) dx_2$$

$$= \Pr\left(C_{[2]} \leq 70\right) + \int_{10}^{x-60} \int_{70-x_2}^{60} \frac{1}{40} dx_1 \frac{1}{20} dx_2$$

$$+ \int_{x-60}^{30} \int_{70-x_2}^{x-x_2} \frac{1}{40} dx_1 \frac{1}{20} dx_2$$

$$f_{C_{[2]}}(x) = \frac{90-x}{800}$$

In summary,

$$f_{C_{[2]}}(x) = \begin{cases} \dfrac{x-30}{800} & 30 \leq x \leq 50 \\[2ex] \dfrac{1}{40} & 50 \leq x \leq 70 \\[2ex] \dfrac{90-x}{800} & 70 \leq x \leq 90 \\[1ex] 0 & \text{otherwise} \end{cases}$$

Next, $C_{[3]} = C_{[2]} + P_{[3]}$. Consider the following partition on the region of support:

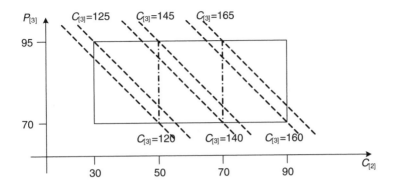

Using similar techniques of analytical integration, we have

$$
f_{C_{[3]}}(x) = \begin{cases}
\dfrac{(x-100)^2}{40,000} & 100 \le x \le 120 \\[2ex]
\dfrac{1}{100} + \dfrac{x-120}{1,000} & 120 \le x \le 125 \\[2ex]
\dfrac{x-120}{1,000} - \dfrac{(x-105)(x-145)}{40,000} & 125 \le x \le 140 \\[2ex]
-\dfrac{(x-105)(x-145)}{40,000} + \dfrac{1}{50} - \dfrac{(x-140)(x-180)}{40,000} & 140 \le x \le 145 \\[2ex]
\dfrac{165-x}{1,000} - \dfrac{(x-140)(x-180)}{40,000} & 145 \le x \le 160 \\[2ex]
\dfrac{1}{100} + \dfrac{165-x}{1,000} & 160 \le x \le 165 \\[2ex]
\dfrac{(x-185)^2}{40,000} & 165 \le x \le 185 \\[2ex]
0 & \text{otherwise}
\end{cases}
$$

Therefore, the expectation and variance of job tardiness can be calculated as

$$
E[T_{[1]}] = \int_{20}^{58} 0 \cdot f_{C_{[1]}}(x) \cdot dx + \int_{58}^{60} (x-58) \cdot f_{C_{[1]}}(x) \cdot dx
$$

$$
= \int_{58}^{60} (x-58) \frac{1}{40} \cdot dx = \frac{1}{20}
$$

$$
E[T_{[1]}^2] = \int_{58}^{60} (x-58)^2 \frac{1}{40} \cdot dx = \frac{1}{15}
$$

$$\text{var}[T_{[1]}] = E\left[T^2_{[1]}\right] - \left(E[T_{[1]}]\right)^2 = \frac{77}{1,200}$$

$$E[T_{[2]}] = \int_{30}^{87} 0 \cdot f_{C_{[2]}}(x) \cdot dx + \int_{87}^{90} (x - 87) \cdot f_{C_{[2]}}(x) \cdot dx$$

$$= \int_{87}^{90} (x - 87) \cdot \frac{90 - x}{800} \cdot dx = \frac{9}{1,600}$$

$$E\left[T^2_{[2]}\right] = \int_{87}^{90} (x - 87)^2 \cdot \frac{90 - x}{800} \cdot dx = \frac{27}{3,200}$$

$$\text{var}[T_{[2]}] = E\left[T^2_{[2]}\right] - \left(E[T_{[2]}]\right)^2 = \frac{21,519}{2,560,000}$$

$$E[T_{[3]}] = \int_{100}^{175} 0 \cdot f_{C_3}(x) \cdot dx + \int_{175}^{185} (x - 175) \cdot f_{C_3}(x) \cdot dx$$

$$= \int_{175}^{185} (x - 175) \cdot \frac{(x - 185)^2}{40,000} \cdot dx = \frac{1}{48}$$

$$E\left[T^2_{[3]}\right] = \int_{175}^{185} (x - 175)^2 \cdot \frac{(x - 185)^2}{40,000} \cdot dx = \frac{1}{12}$$

$$\text{var}[T_{[3]}] = E\left[T^2_{[3]}\right] - \left(E[T_{[3]}]\right)^2 = \frac{191}{2,304}$$

To obtain variance of the total tardiness, we apply Equation (4.19) to derive covariance terms:

$$E[T_{[1]}T_{[2]}] = \int_{58}^{60} (x_1 - 58) \int_{87-x_1}^{30} (x_1 + x_2 - 87) f_{P_{[2]}}(x_2) \cdot dx_2 f_{P_{[1]}}(x_1) \cdot dx_1$$

$$= \int_{58}^{60} (x_1 - 58) \int_{87-x_1}^{30} (x_1 + x_2 - 87) \frac{1}{20} \cdot dx_2 \frac{1}{40} \cdot dx_1 = \frac{17}{2,400}$$

$$\text{cov}[T_{[1]}, T_{[2]}] = E[T_{[1]}T_{[2]}] - E[T_{[1]}]E[T_{[2]}] = \frac{653}{96,000}$$

$$E[T_{[2]}T_{[3]}] = \int_{87}^{90} (c_2 - 87) \int_{175-c_2}^{95} (c_2 + x_3 - 175) f_{P_{[3]}}(x_3) \cdot dx_3 f_{C_{[2]}}(c_2) \cdot dc_2$$

$$= \int_{87}^{90} (c_2 - 87) \int_{175-c_2}^{95} (c_2 + x_3 - 175) \frac{1}{25} \cdot dx_3 \frac{90 - c_2}{800} \cdot dc_2$$

$$= \frac{6,543}{800,000}$$

$$\text{cov}[T_{[2]}, T_{[3]}] = E[T_{[2]}T_{[3]}] - E[T_{[2]}]E[T_{[3]}] = \frac{25,797}{3,200,000}$$

Define $P_{[2]+[3]} = P_{[2]} + P_{[3]}$ so that $T_{[3]} = \max\{P_{[1]} + P_{[2]+[3]} - d_{[3]}, 0\}$

Using an approach similar to the derivation of $C_{[2]} = P_{[1]} + P_{[2]}$, we have

$$
f_{P_{[2]+[3]}}(x) = \begin{cases} \dfrac{x - 80}{500} & 80 \leq x \leq 100 \\[2mm] \dfrac{1}{25} & 100 \leq x \leq 105 \\[2mm] \dfrac{125 - x}{500} & 105 \leq x \leq 125 \\[2mm] 0 & \text{otherwise} \end{cases}
$$

Hence

$$
\begin{aligned}
& E[T_{[1]}T_{[3]}] \\
&= \int_{58}^{60} (x_1 - 58) \int_{175-x_1}^{125} (x_1 + x - 175) f_{P_{[2]+[3]}}(x) \cdot dx \cdot f_{P_{[i]}}(x_1) \cdot dx_1 \\
&= \int_{58}^{60} (x_1 - 58) \int_{175-x_1}^{125} (x_1 + x - 175) \frac{125 - x}{500} \cdot dx \cdot \frac{1}{40} \cdot dx_1 \\
&= \frac{128}{9,375}
\end{aligned}
$$

$$
\text{cov}[T_{[1]}, T_{[3]}] = E[T_{[1]}T_{[3]}] - E[T_{[1]}] E[T_{[3]}] = \frac{7,567}{600,000}
$$

Finally, we have

$$
E\left[\sum_{j=1}^{n} T_{[j]}\right] = \sum_{j=1}^{n} E[T_{[j]}] = \frac{367}{4,800} \approx 0.07646
$$

$$
\begin{aligned}
\text{var}\left[\sum_{j=1}^{n} T_{[j]}\right] &= \sum_{j=1}^{n} \text{var}[T_{[j]}] + 2\sum_{i=1}^{n-1} \sum_{j=i+1}^{n} \text{cov}[T_{[i]}, T_{[j]}] \\
&= \frac{24,240,667}{115,200,000} \approx 0.21042.
\end{aligned}
$$

A.2 Analysis for a Single-Machine Total Number of Tardy Jobs Problem

Job Number	1	2	3
Processing time	$U(20, 60)$	$U(10, 30)$	$U(70, 95)$
Due date	58	87	175

Consider the processing sequence 1–2–3. The completion time distributions were derived in Section A.1. Therefore, we have

$$E[U_{[1]}] = \int_{58}^{60} f_{C_{[1]}}(x) \cdot dx = \int_{58}^{60} \frac{1}{40} \cdot dx = \frac{1}{20}$$

$$\mathrm{var}[U_{[1]}] = E[U_{[1]}] \cdot (1 - E[U_{[1]}]) = \frac{19}{400}$$

$$E[U_{[2]}] = \int_{87}^{90} f_{C_{[2]}}(x) \cdot dx = \int_{87}^{90} \frac{90 - x}{800} \cdot dx = \frac{9}{1,600}$$

$$\mathrm{var}[U_{[2]}] = E[U_{[2]}] \cdot (1 - E[U_{[2]}]) = \frac{14,319}{2,560,000}$$

$$E[U_{[3]}] = \int_{175}^{185} f_{C_{[3]}}(x) \cdot dx = \int_{175}^{185} \frac{(x - 185)^2}{40,000} \cdot dx = \frac{1}{120}$$

$$\mathrm{var}[U_{[3]}] = E[U_{[3]}] \cdot (1 - E[U_{[3]}]) = \frac{119}{14,400}$$

Furthermore, we have

$$E[U_{[1]}U_{[2]}] = \int_{58}^{60} \int_{87-x_1}^{30} f_{P_{[2]}}(x_2) \cdot dx_2 f_{P_{[1]}}(x_1) \cdot dx_1$$

$$= \int_{58}^{60} \int_{87-x_1}^{30} \frac{1}{20} \cdot dx_2 \frac{1}{40} \cdot dx_1 = \frac{1}{200}$$

$$\mathrm{cov}[U_{[1]}, U_{[2]}] = E[U_{[1]}U_{[2]}] - E[U_{[1]}] E[U_{[2]}] = \frac{151}{32,000}$$

$$E[U_{[2]}U_{[3]}] = \int_{87}^{90} \int_{175-c_2}^{95} f_{P_{[3]}}(x_3) \cdot dx_3 f_{C_{[2]}}(c_2) \cdot dc_2$$

$$= \int_{87}^{90} \int_{175-c_2}^{95} \frac{1}{25} \cdot dx_3 \frac{90 - c_2}{800} \cdot dc_2 = \frac{9}{5,000}$$

$$\mathrm{cov}[U_{[2]}, U_{[3]}] = E[U_{[2]}U_{[3]}] - E[U_{[2]}] E[U_{[3]}] = \frac{561}{320,000}$$

$$E[U_{[1]}U_{[3]}] = \int_{58}^{60} \int_{175-x_1}^{125} f_{P_{[2]+[3]}}(x) \cdot dx \cdot f_{P_{[1]}}(x_1) \cdot dx_1$$

$$= \int_{58}^{60} \int_{175-x_1}^{125} \frac{125 - x}{500} \cdot dx \cdot \frac{1}{40} \cdot dx_1 = \frac{61}{15,000}$$

$$\mathrm{cov}[U_{[1]}, U_{[3]}] = E[U_{[1]}U_{[3]}] - E[U_{[1]}] E[U_{[3]}] = \frac{73}{20,000}$$

Hence

$$E\left[\sum_{j=1}^{n} U_{[j]}\right] = \sum_{j=1}^{n} E\left[U_{[j]}\right] = \frac{307}{4800} \approx 0.06396$$

$$\operatorname{var}\left[\sum_{j=1}^{n} U_{[j]}\right] = \sum_{j=1}^{n} \operatorname{var}\left[U_{[j]}\right] + 2\sum_{i=1}^{n-1}\sum_{j=i+1}^{n} \operatorname{cov}\left[U_{[i]}, U_{[j]}\right]$$

$$= \frac{1{,}880{,}087}{23{,}040{,}000} \approx 0.08160$$

A.3 Analysis for a Single-Machine Maximum Lateness Problem

Job Number	1	2	3
Processing time	$U(20, 60)$	$U(10, 30)$	$U(70, 95)$
Due date	58	87	175

Consider the processing sequence 1–2–3. Note that

$$L_{[1]} = P_{[1]} - d_{[1]} = P_{[1]} - 58$$
$$L_{[2]} = P_{[1]} + P_{[2]} - d_{[2]} = P_{[1]} + P_{[2]} - 87$$
$$L_{[3]} = P_{[1]} + P_{[2]} + P_{[3]} - d_{[3]} = P_{[1]} + P_{[2]} + P_{[3]} - 175$$

Hence

$$L_{[1]} \geq L_{[2]} \Leftrightarrow P_{[1]} - 58 \geq P_{[1]} + P_{[2]} - 87 \Leftrightarrow P_{[2]} \leq 29$$
$$L_{[2]} \geq L_{[3]} \Leftrightarrow P_{[1]} + P_{[2]} - 87 \geq P_{[1]} + P_{[2]} + P_{[3]} - 175 \Leftrightarrow P_{[3]} \leq 88$$
$$L_{[3]} \geq L_{[1]} \Leftrightarrow P_{[1]} + P_{[2]} + P_{[3]} - 175 \geq P_{[1]} - 58 \Leftrightarrow P_{[2]} + P_{[3]} \geq 117$$

To determine the maximum among the job lateness values, we used the following partition:

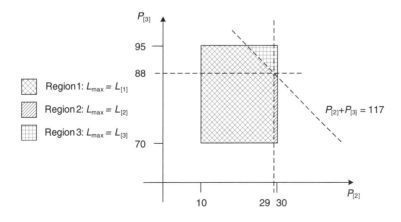

Therefore,

$$
\begin{aligned}
E[L_{\max}] &= \int_{20}^{60} f_{P_1}(x_1) \cdot \left(\int_{10}^{22} \int_{70}^{95} (x_1 - 58) f_{P_{[3]}}(x_3) \cdot dx_3 f_{P_{[2]}}(x_2) \cdot dx_2 \right. \\
&\quad + \int_{22}^{29} \int_{70}^{117-x_2} (x_1 - 58) f_{P_{[3]}}(x_3) \cdot dx_3 f_{P_{[2]}}(x_2) \cdot dx_2 \quad \text{(Region 1)} \\
&\quad + \int_{29}^{30} \int_{70}^{88} (x_1 + x_2 - 87) f_{P_{[3]}}(x_3) \cdot dx_3 f_{P_{[2]}}(x_2) \cdot dx_2 \quad \text{(Region 2)} \\
&\quad + \int_{88}^{95} \int_{117-x_3}^{30} (x_1 + x_2 + x_3 - 175) f_{P_{[2]}}(x_2) \cdot dx_2 \\
&\quad \quad \left. \times f_{P_{[3]}}(x_3) \cdot dx_3 \right) \cdot dx_1 \quad \text{(Region 3)} \\
&= -\frac{10687}{600} \approx -17.81167
\end{aligned}
$$

$$
\begin{aligned}
E\left[L_{\max}^2\right] &= \int_{20}^{60} f_{P_1}(x_1) \cdot \left(\int_{10}^{22} \int_{70}^{95} (x_1 - 58)^2 f_{P_{[3]}}(x_3) \cdot dx_3 f_{P_{[2]}}(x_2) \cdot dx_2 \right. \\
&\quad + \int_{22}^{29} \int_{70}^{117-x_2} (x_1 - 58)^2 f_{P_{[3]}}(x_3) \cdot dx_3 f_{P_{[2]}}(x_2) \cdot dx_2 \quad \text{(Region 1)} \\
&\quad + \int_{29}^{30} \int_{70}^{88} (x_1 + x_2 - 87)^2 f_{P_{[3]}}(x_3) \cdot dx_3 f_{P_{[2]}}(x_2) \cdot dx_2 \quad \text{(Region 2)} \\
&\quad + \int_{88}^{95} \int_{117-x_3}^{30} (x_1 + x_2 + x_3 - 175)^2 f_{P_{[2]}}(x_2) \cdot dx_2 \\
&\quad \quad \left. \times f_{P_{[3]}}(x_3) \cdot dx_3 \right) \cdot dx_1 \quad \text{(Region 3)} \\
&= \frac{2{,}707{,}487}{6000}
\end{aligned}
$$

$$
\text{var}[L_{\max}] = E\left[L_{\max}^2\right] - E[L_{\max}]^2 = \frac{48{,}237{,}251}{360{,}000} \approx 133.99236
$$

A.4 Software XVA-Sched

XVA-Sched is a user-friendly software that helps in implementing the methodologies developed in Chapters 4 through 8 to determine expectation-variance (EV)–efficient schedules. As discussed in these chapters, these schedules pertain to diverse machine configurations and appropriate performance measures. XVA-Sched is developed for Windows XP platform, and it requires installation of MATLAB Compiler Runtime (version 7.8) for its use. Next, we present the procedure to use this software.

A.4.1 Starting the Software

The following welcome screen appears when the user accesses the software (Figure A.1).

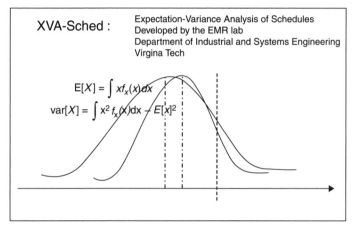

Figure A.1. Welcome screen of XVA-Sched.

After the software is initialized, the welcome screen disappears, and the user is asked to specify the basic configurations of the problem. The first step in that regard is the selection of machine configuration, as shown in Figure A.2.

This is followed by selection of a performance measure. The choices available to the user for the selection of a performance measure depend on the machine configuration selected in the first step. These options are shown in Table A.1.

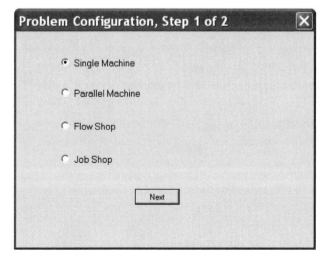

Figure A.2. Selection of machine configuration.

Table A.1. *Selection of Performance Measure*

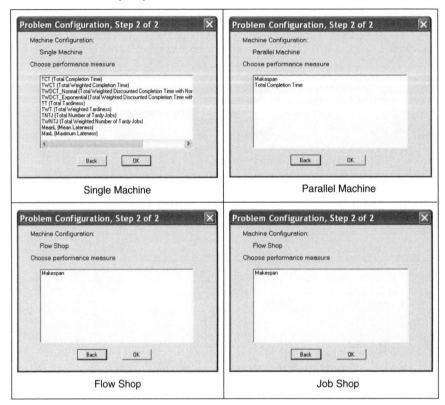

Single Machine	Parallel Machine
Flow Shop	Job Shop

Note that the user can go back to the first step by clicking on the "Back" button if need be. To confirm selection of the problem configuration, the user must click on the "OK" button. The software brings up the main interface once the "OK" button is clicked. The exact appearance of this interface may vary depending on the selected problem configuration, but the basic layout is the same, as shown in Figure A.3. Note that after the initial configuration is confirmed, the user still can switch from one configuration to another by using the machine configuration tabs that are located under the menu bar.

Next, we describe the procedures for performing the EV analyses for single-machine, parallel-machine, flow-shop, and job-shop configurations, respectively.

A.4.2 Single Machine

The layout for single-machine configuration is shown in Figure A.3. Suppose that we have selected total weighted tardiness as the performance measure.

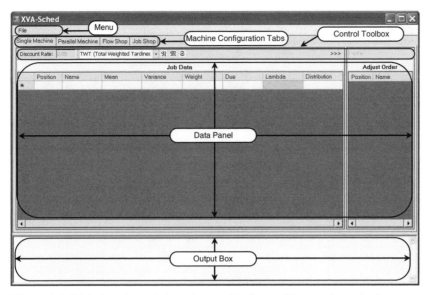

Figure A.3. Main interface of the software (single-machine configuration).

Step 1. Input job data.

On the selection of this machine configuration, the following table appears on the left side of the data panel:

	Position	Name	Mean	Variance	Weight	Due	Lambda	Distribution
*								

Each row of this table corresponds to a job. The columns pertain to the following information about each job:

Position: Position of the job in a processing sequence

Name: Name of the job

Mean: Expected processing time of the job

Variance: Variance of processing time for the job

Weight: Importance factor of the job in a performance measure

Due: Job due date

Lambda: Processing rate, when the job processing time is assumed to be exponentially distributed for the calculation of the total weighted discounted completion time

Distribution: Processing time distribution of the job (This is used for mixture analysis and simulation approaches; also see Section A.4.5.)

Note that the user needs to provide information only for the columns that are relevant for the selected performance measure. A column for which information is not required will be shaded gray. For example, the value of lambda is not relevant to calculation of total weighted tardiness. Consequently, the column corresponding to lambda will appear shaded gray as shown below:

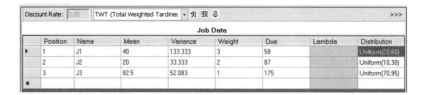

Step 2. Adjust job sequence.

Besides directly inputting job positions in the preceding data table, the user also can adjust job sequence by using the table on the right side of the data panel. To do this, the user needs to click on the following button in the control toolbox:

.

The table on the right will depict job names, sorted according to their positions in the processing sequence, as shown below:

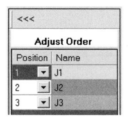

Suppose that the current sequence is J1–J2–J3. To change the sequence and place job J2 at the end, the user can access the drop-down selector next to J2 and pick 3 as its desired position.

 →

After reordering these jobs, the user needs to click the following update button to register the new sequence back into the data table on the left side of the data panel:

<<<
.

Step 3. Select evaluation approach.

Finally, to perform the expectation and variance analysis, the user needs to choose from the following three approaches:

$\mathfrak{N}$: Normal approximation
$\mathfrak{M}$: Mixture approximation
$\mathfrak{S}$: Monte Carlo simulation

Note:

For some of the listed performance measures, the approach of normal approximation can be used to obtain exact results of expectation and variance. Hence the first approach is denoted as $\mathfrak{A}$ (for "analytical results"). The other two approaches are not necessary and are not offered as options for those cases.

If the approach of mixture approximation is selected, the user will be prompted to input the (common) number of components to be used to approximate the processing time distribution for each job.

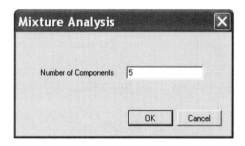

The software outputs the results at the bottom of the interface. It provides information on the run number of the session; the input provided by the user on machine configuration, number of jobs, processing sequence, and approach used; and the values of the expectation and variance of the selected performance measure for the processing sequence as follows:

```
{ No. 1}======================================
Machine configuration:  [Single Machine]
Number of Jobs: [ 3 ]
Processing sequence:    [ J1 -> J2 -> J3 ]
Approach used:  [Normal Approximation]
Expectation of  [MaxL] = [-17.65 ]
Variance of     [MaxL] = [ 135.45 ]
-----------------------------------------------
```

Step 4. Change performance measure.

Besides using the initially selected performance measure, the user also can pick a different performance measure from the drop-down list in the control toolbox.

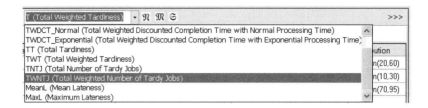

Note that the user may need to input other pertinent information for the jobs before he or she can perform expectation and variance analysis with the newly selected performance measure.

A.4.3 Parallel Machine

The layout for a parallel-machine configuration is shown in Figure A.4.

For this case, the data panel is split into three parts. The table in the middle shows the list of machines and also the processing sequence on each machine. The table on the left is used to input job-related data. The table on the right is used to adjust processing sequence on a chosen machine.

Step 1. Specify number of machines.

Input the number of machines in the following box, and press "Setup":

This will populate the sequence table with the desired number of rows, with each row corresponding to a machine. The machines are automatically named

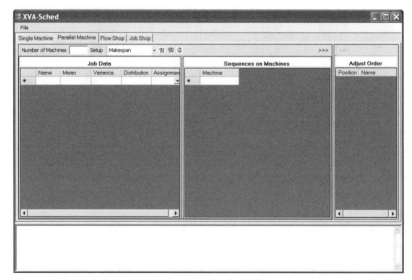

Figure A.4. Main interface of the software (parallel-machine configuration).

as "Machine n," $n = 1, 2, \ldots$. The user has the option of changing the names of the machines to any appropriate nominals.

Step 2. Input job data.
The table for job data is on the left side of the data panel. The columns pertain to the following information about each job:

Name: Name of the job

Mean: Expected processing time of the job

Variance: Variance of processing time for the job

Distribution: Processing time distribution of the job (This is used for mixture analysis and simulation; also see Section A.4.5.)

Assignment: The name of the machine to which this job is assigned

Note that the "Assignment" column contains a drop-down list for each job, the content of which reflects the names of the machines. The user needs to input a machine name in the middle table first before any job can be assigned to it. If a machine is deleted, then the jobs assigned to it are also deleted. When the assignment of a job is changed from "Machine i" to "Machine j," it is removed from the processing sequence on "Machine i" and inserted at the end of processing sequence on "Machine j."

Step 3. Adjust processing sequence.
To adjust the processing sequence on a particular machine, the user first needs to select the corresponding row in the sequence table and then click on the following button:

.

The table on the right will depict job names, sorted according to their positions in the processing sequence. Suppose that the sequence on the current machine is J1–J2–J3:

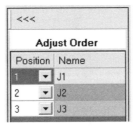

To change this sequence and place job J2 at the end, the user can access the drop-down selector next to J2 and pick 3 as its desired position:

 →

After reordering the jobs, the user needs to click the following update button:

.

This registers the new sequence back into the selected row in the sequence table:

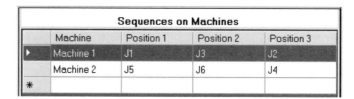

Step 4. Select evaluation approach.

For the parallel-machine configuration, there are two supported performance measures: makespan and total completion time. The user needs to choose from one of the following approaches to calculate the expectation and variance of the performance measure:

  $\mathfrak{N}$: Normal approximation
  $\mathfrak{M}$: Mixture approximation
  $\mathfrak{S}$: Monte Carlo simulation

If the approach of mixture approximation is selected, the user will be prompted to input the (common) number of components to be used to approximate the processing time distribution of each job.

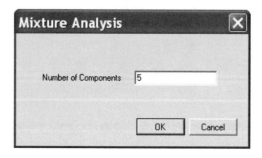

The software outputs the results at the bottom of the interface. It provides information on the run number of the session; the input provided by the user on machine configuration, number of jobs, number of machines, processing sequence, and approach used; and the values of the expectation and variance of the selected performance measure for the processing sequence as follows:

```
{ No. 2}=======================================
Machine configuration: [Parallel Machine]
Number of jobs:      [ 6 ]
Number of machines:      [ 2 ]
Processing sequence:
    [ J1 -> J3 -> J6 -> J4 ]
    [ J2 -> J5 ]
Approach used:   [Normal Approximation]
Expectation of   [Makespan] = [ 23.607 ]
Variance of      [Makespan] = [ 8.4 ]
-------------------------------------------------
```

A.4.4 Flow Shop

The layout for flow-shop configuration is shown in Figure A.5.

Step 1. Specify job routing and processing sequence.
Enter the sequence in which the machines are to be visited ("Machine List") for each job and the sequence in which the jobs are to be processed on each machine ("Job List").

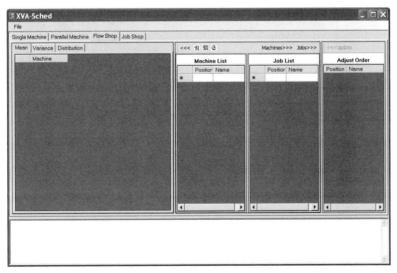

Figure A.5. Main interface of the software (flow-shop configuration).

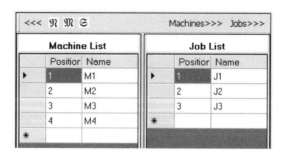

Note that the job routing ("Machine List") and the processing sequence ("Job List") can be adjusted, respectively, by clicking on one of the following buttons:

.

Suppose that the current processing sequence is J1–J2–J3, and the user has clicked the "Jobs" button; the table on the right (named "Adjust Order") will depict job names sorted according to their positions in the processing sequence.

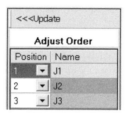

To change the sequence and place job J2 at the end, the user can access the drop-down selector next to J2 and pick 3 as its desired position:

After reordering the jobs, the user needs to click the following update button:

.

This registers the new sequence back into the job list. Adjustment of job routing can be performed in a similar manner.

Step 2. Input processing time data.

Click the following button

,

so that data tables on the left side of the panel can be initialized with appropriate numbers of columns and rows:

Mean	Variance	Distribution		
	Machine	J1	J2	J3
▶	M1			
	M2			
	M3			
	M4			

Note that three data tables are provided to specify expectation, variance, and distribution of job processing times, respectively. Each row in a data table corresponds to a particular machine, and each column corresponds to a particular job. The distribution data table is only relevant if the user plans to use mixture approximation or Monte Carlo simulation as the evaluation approach in step 3.

Step 3. Select evaluation approach.

The user needs to choose from the following three approaches to calculate the expectation and variance of the makespan:

$\mathfrak{N}$: Normal approximation
$\mathfrak{M}$: Mixture approximation
$\mathfrak{S}$: Monte Carlo simulation

If the approach of mixture approximation is selected, the user will be prompted to input the (common) number of components to be used to approximate the processing time distribution of each job on each machine:

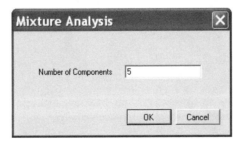

The user also needs to specify the maximum number of mixture components to be used at intermediate steps. If the number of components exceeds this limit, a mixture reduction procedure will be applied automatically to reduce the number of components:

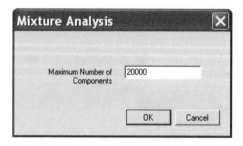

The software outputs the results at the bottom of the interface. It provides information on the run number of the session; the input provided by the user on machine configuration, number of jobs, number of machines, job routing, processing sequence, and approach used; and the values of the expectation and variance of the selected performance measure for the processing sequence as follows:

```
{ No. 3}=======================================
Machine configuration:  [Flow Shop]
Number of jobs: [ 4 ]
Number of machines: [ 4 ]
Job routing:     [ M1 -> M2 -> M3 -> M4 ]
Processing sequence:    [ J2 -> J4 -> J3 -> J1 ]
Approach used:  [Normal Approximation]
Expectation of  [Makespan] = [ 38.485 ]
Variance of     [Makespan] = [ 12.145 ]
-----------------------------------------------
```

A.4.5 Job Shop

The layout for job-shop configuration is shown in Figure A.6.

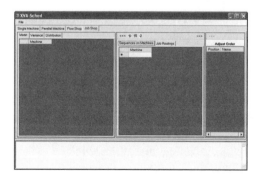

Figure A.6. Main interface of the software (job-shop configuration).

Step 1. Specify job routing and processing sequence.

Use the sequence table in the middle of data panel to input names of machines:

Sequences on Machines	Job Routings
Machine	
M1	
M2	
M3	
▸*	

Click on the "Job Routings" tab to edit the routing table:

Job	Step 1	Step 2	Step 3
J1			
J2			
J3			
*			

Sequences on Machines | Job Routings

Note that the number of columns in the routing table increases with the number of machines specified in the sequence table. After the user inputs all job routings, he or she needs to go back to the sequence table to specify processing sequence on each machine. (The number of columns in the sequence table also increases with the number of jobs in the routing table.)

Sequences on Machines | Job Routings

Machine	Position 1	Position 2	Position 3
M1	J1	J2	J3
M2			
M3			
*			

The routing of a job or processing sequence on a machine can be adjusted by clicking on the following button when the corresponding row is selected in the two tables introduced above:

Suppose that the user has selected machine M1, on which the processing sequence is J1–J2–J3. The table on the right (named "Adjust Order") will depict job names sorted according to their positions in the processing sequence:

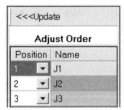

To change the sequence and place job J2 at the end, the user can access the drop-down selector next to J2 and pick 3 as its desired position:

 →

After reordering these jobs, the user needs to click the following update button:

 .

This registers the new sequence back into the sequence table. Adjustment of job routing can be performed in a similar manner.

Step 2. Input processing time data.
Click on the following button

 ,

so that data tables on the left side of the panel can be initialized with appropriate numbers of columns and rows:

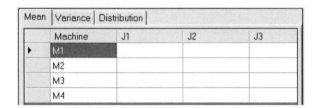

Note that three data tables are provided to specify expectation, variance, and distribution of job processing times, respectively. Each row in a data table corresponds to a particular machine, and each column corresponds to a particular job. The distribution data table is relevant only if the user plans to use mixture approximation or Monte Carlo simulation as the evaluation approach in step 3.

Step 3. Select evaluation approach.

The user needs to choose from the following three approaches to calculate the expectation and variance of the makespan:

$\mathfrak{N}$: Normal approximation
$\mathfrak{M}$: Mixture approximation
$\mathfrak{S}$: Monte Carlo simulation

If the approach of mixture approximation is selected, the user will be prompted to input the (common) number of components to be used to approximate the processing time distribution of each job on each machine:

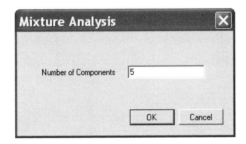

The user also needs to specify the maximum number of mixture components to be used at intermediate steps. If the number of components exceeds this limit, a mixture reduction procedure will be applied automatically to reduce the number of components:

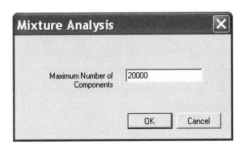

The software outputs the results at the bottom of the interface. It provides information on the run number of the session; the input provided by the user on machine configuration, number of jobs, number of machines, and approach used; and the values of the expectation and variance of the selected performance measure for the processing sequence as follows:

```
{ No. 4}======================================
Machine configuration:   [Job Shop]
Number of jobs: [ 3 ]
Number of machines: [ 3 ]
Approach used:   [Normal Approximation]
Expectation of   [Makespan] = [ 177.86 ]
Variance of      [Makespan] = [ 68.507 ]
```

A.4.6 Specifying Distributions

By double-clicking on any data cell that is attributed to the distribution of processing time, the following dialog box will be displayed to enable the user to input relevant data:

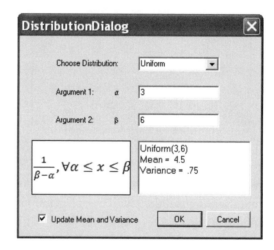

Note that when a particular type of distribution is selected, the names of its arguments and the corresponding density function will be updated to reflect this change. If the values of the arguments are changed, the mean and variance of the distribution will be calculated and reported in the text box at the lower right corner:

Uniform(3,6)
Mean = 4.5
Variance = .75

The checkbox at the bottom

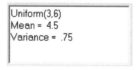
,

is included to confirm whether the user wants to update the mean and variance values in the corresponding data table.

The following distributions have been incorporated into the program:

Normal
Uniform
Exponential
Rayleigh
LogNormal
Weibull
Beta
Gamma
Empirical

For the first four distribution options, mixture components have been precalculated for mixtures with less than or equal to 5 components to speed up the calculations. If the last option, "Empirical," is chosen, the user needs to press the "Edit" button

,

to enter the empirical probability density function in the following dialog box:

To illustrate how an empirical distribution is specified in the table, the density function corresponding to the data shown in Figure A.7 is plotted in Figure A.8.

Besides entering the values directly, the user also can load the CDF from a CSV (comma separated values) file by clicking on the button

.

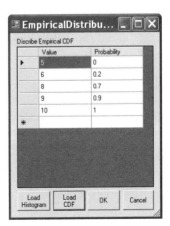

Figure A.7. Empirical distribution dialog box.

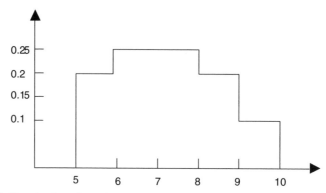

Figure A.8. Density function of an empirical distribution.

As an alternative, the user also can load the histogram from a CSV file by using the button

.

Loading a histogram file with the following content leads to the same empirical distribution as that shown in Figure A.6.

5	0
6	200
8	500
9	200
10	100

Note that the lower bound of the first data bin is always zero. The bins and frequencies for the preceding histogram are

Bin	[0, 5]	[5, 6]	[6,8]	[8, 9]	[9, 10]
Frequency	0	200	500	200	100

Bibliography

Agrawal, M. K., and Elmaghraby, S. E. 2001. "On computing the distribution function of the sum of independent random variables." *Computers and Operations Research* 28(5):473–483.

Ayhan, H., and Olsen, T. L. 2000. "Scheduling of multiclass single server queues under nontraditional performance measures." *Operations Research* 48(3):482–489.

Baker, K. R. 1974. *Introduction to Sequencing and Scheduling.* New York: Wiley.

Clark, C. E. 1961. "The greatest of a finite set of random variables." *Operations Research* 9:145–162.

Daniels, R. L., and Chambers, R. J. 1990. "Multiobjective flow-shop scheduling." *Naval Research Logistics* 37:981–995.

Daniels, R. L., and Kouvelis, P. 1995. "Robust scheduling to hedge against processing time uncertainty in single-stage production." *Management Science* 41(2):363–376.

Daniels, R. L., and Carrillo, J. E. 1997. "β-Robust scheduling for single-machine systems with uncertain processing times." *IIE Transactions* 29:977–985.

Dar-el, E. M., Herer, Y. T., and Masin, M. 1999. "CONWIP-based production lines with multiple bottlenecks: Performance and design implications." *IIE Transactions* 31:99–111.

De, P., Ghosh, J. B., and Wells, C. E. 1992. "Expectation-variance analysis of job sequences under processing time uncertainty." *International Journal of Production Economics* 28:289–297.

Dempster, A. P., Laird, N. M., and Rubin, D. B. 1977. "Maximum likelihood from incomplete data via the EM algorithm." *Journal of the Royal Statistical Society, Series B (Methodological)* 39(1):1–38.

de Kluyver, C. A. 1980. "Media selection by mean-variance analysis." *European Journal of Operational Research* 5:112–117.

de Kluyver, C. A., and Baird, F. T. 1984. "Media selection by mean-variance analysis." *European Journal of Operational Research* 16:152–156.

Dodin, B. 1985. "Approximating the distribution function in stochastic networks." *Computers and Operations Research* 12(3):251–264.

Dodin, B. 1996. "Determining the optimal sequences and the distributional properties of the completion times in stochastic flow shops." *Computers and Operations Research* 23:829–843.

Duenyas, I., Hopp, W. H., and Spearman, M. L. 1993. "Characterizing the output process of a CONWIP line with deterministic processing and random outages." *Management Science* 39:975–988.

Elmaghraby, S. E. 1977. *Activity Networks: Project Planning and Control by Network Models*. New York: Wiley.

Forst, F. G. 1995. "Bicriteria stochastic scheduling on one or more machines." *European Journal of Operational Research* 80:404–409.

Glover, F. 1975. "Surrogate constraint duality in mathematical programming." *Operations Research* 23:434–453.

Held, M., and Karp, R. M. 1962. "A dynamic programming approach to sequencing problems." *The Japan Society for Industrial and Applied Mathematics (SIAM)* 10:196–210.

Hendricks, K. B. 1992. "The output processes of serial production lines of exponential machines with finite buffers." *Operations Research* 40(6):1139–1147.

Johnson, S. M. 1954. "Optimal two- and three-stage production schedules with setup times included." *Naval Research Logistics Quarterly* 1:61–67.

Jung, Y. S., Nagasawa, H., and Nishiyama, N. 1990. "Bicriteria single-stage scheduling to minimize both the expected value and the variance of the total flow time." *Journal of Japan Industrial Management Association* 39:76–82 (in Japanese).

Kouvelis, P., Daniels, R. L., and Vairaktarakis, G. 2000. "Robust scheduling of a two-machine flow shop with uncertain processing times." *IIE Transactions* 32:421–432.

Kumar, S., and Kumar, P. R. 1994. "Fluctuation smoothing policies are stable for stochastic reentrant lines." In: *33rd IEEE Proceedings Conference on Decision and Control*, December, pp. 1476–1480.

Lin, C., and Lee, C. 1995. "Single-machine stochastic scheduling with dual criteria." *IIE Transactions* 27:244–249.

Lin, K. S. 1983. "Hybrid algorithm for sequencing with bicriteria." *Journal of Optimization Theory and Applications* 39(1):105–124.

Liu, Q., Ohno, K., and Nakayama, H. 1992. "Multiobjective discounted Markov decision processes with expectation and variance criteria." *International Journal of Systems Science* 23(6):903–914.

Lu, S. C. H., Ramaswamy, D., and Kumar, P. R. 1994. "Efficient scheduling policies to reduce mean and variance of cycle-time in semiconductor manufacturing plants." *IEEE Transactions on Semiconductor Manufacturing* 7(3):374–385.

Mazzola, J. B., and Neebe, A. W. 1986. "Resource-constrained assignment scheduling." *Operations Research* 34:560–572.

McKay, K. N., Safayeni, F. R., and Buzacott, J. A. 1988. "Job-shop scheduling theory: What is relevant?" *Interfaces* 18:84–90.

McLachlan, G. J., and Peel, D. 2001. *Finite Mixture Models*. New York: Wiley.

Mclachlan, G. J., and Krishnan, T. 1997. *The EM Algorithm and Extensions*. New York: Wiley.

Morizawa, K., Ono, T., Nagasawa H., and Nishiyama, N. 1993. "An interactive approach for searching a preferred schedule." *Journal of Japan Industrial Management Association* 39:76–82.

Murata, T., Ishibuchi, H., and Tanaka, H. 1996. "Multiobjective genetic algorithm and its applications to flowshop scheduling." *Computers and Industrial Engineering* 30:957–968.

Nagasawa, H., and Shing, C. 1997. "Interactive decision system in parallel-machine stochastic multi-objective scheduling." In: *Proceedings of the 1st International Conference on Engineering Design and Automation*, Bangkok, Thailand, pp. 421–424.

Nagasawa, H., and Shing, C. 1998. "Interactive decision system in stochastic multi-objective scheduling to minimize the expected value and variance of total flow time." *Journal of the Operations Research Society of Japan* 41(2):261–278.

Nelson, R. T., Sarin, R. K., and Daniels, R. L. 1986. "Scheduling with multiple performance measures: The one-machine case." *Management Science* 32(4):464–479.

Pinedo, M. 2002. *Scheduling: Theory, Algorithms, and Systems*, 2nd ed. Englewood Cliffs, NJ: Prentice-Hall.

Portougal, V., and Trietsch, D. 1998. "Makespan related criteria for comparing schedules in stochastic environments." *The Journal of the Operational Research Society*. 49(11):1188–1195.

Rajendran, C. 1995. "Heuristics for scheduling in flowshop with multiple objectives." *European Journal of Operational Research* 82:540–555.

Runnalls, A. R. 2007. "Kullback-Leibler approach to Gaussian mixture reduction." *IEEE Transactions on Aerospace and Electronic Systems* 43(3):989–999.

Salmond, D. J. 1988. "Mixture reduction algorithms for uncertainty tracking." Technical Report 88004. Royal Aerospace Establishment, Farnborough, England.

Sarin, S. C., and Hariharan, R. 2000. "A two machine bicriteria scheduling problem." *International Journal of Production Economics* 65:125–139.

Sarin, S. C., and Prakash, D. 2004. "Equal processing time bicriteria scheduling on parallel machines." *Journal of Combinatorial Optimization* 8:227–240.

Sculli, D. 1983. "The completion time of PERT networks." *Journal of the Operational Research Society* 34(2):155–158.

Shayman, M. A., and Gaucherand, E. F. 2001. "Risk-sensitive decision theoretic diagnosis." *IEEE Transactions on Automatic Control* 46(7):1167–1171.

Shing, C., and Nagasawa, H. 1997. "Interactive decision support system in stochastic multi-objective scheduling." *Bulletin of Osaka Prefecture University*, Series A, 45(2):133–142.

Shing, C., and Nagasawa, H. 1999. "Interactive decision support system in stochastic multi-objective portfolio selection." *International Journal of Production Economics* 60–61:187–193.

Shogan, A. 1977. "Bounding distributions for a stochastic PERT network." *Networks* 7(4):359–381.

Spearman, M. L., and Hopp, W. H. 1991. "Throughput of a constant work in process manufacturing line subject to failures." *International Journal of Production Research* 29:635–655.

Spearman, M. L., Woodruff, D. L., and Hopp, W. H. 1980. "CONWIP: A pull alternative to kanban." *International Journal of Production Research* 28(5):879–894.

Soroush, H. M. 1994. "The most critical path in a PERT network: A heuristic approach." *European Journal of Operational Research* 78(1):93–105.

Soroush, H. M., and Fredenhall, L. D. 1984. "The stochastic single machine scheduling problem with earliness and tardiness costs." *European Journal of Operational Research* 77:187–302.

Srivastava, R. K., and Sarin, S. C. 1993. "Determination of part-delivery dates in a small lot stochastic assembly system." *Opsearch* 30(4):281–312.

Tan, B. 1997. "Variance of the throughput of an n-station production line with no intermediate buffers and time dependent failures." *European Journal of Operational Research* 101:560–576.

Tan, B. 2000. "Asymptotic variance rate of the output in production lines with finite buffers." *Annals of Operations Research* 93:385–403.

T'kindt, V., and Billaut, J.-C. 2005. "Special issue on multi-criteria scheduling." *European Journal of Operations Research* 167:796–809.

van de Geer, S. A. 1996. "Rates of convergence for the maximum likelihood estimator in mixture models." *Journal Nonparametric Statistics* 6:293–310.

van de Geer, S. A. 2004. "Asymptotic theory for maximum likelihood in nonparametric mixture models." *Computational Statistics & Data Analysis* 41:453–464.

Wang, B., and Mazumder, P. 2005. "Multivariate normal distribution based statistical timing analysis using global projection and local expansion." In: *Proceedings of the 18th International Conference on VLSI Design,* pp. 380–385.

Wassenhove, V., and Gelders, L. F. 1980. "Solving a bi-criteria scheduling problem." *European Journal of Operational Research* 4(1):42–48.

Wilhelm, W. E. 1986. "A model to approximate transient performance of the flowshop." *International Journal of Production Research* 24(1):33–50.

Wilhelm, W. E. 1986. "The application of lognormal models of transient operations in the flexible manufacturing environment." *Journal of Manufacturing Systems* 5(4): 253–266.

Wilhelm, W. E., and Ahmadi-Marandi, S. 1982. "A methodology to describe operating characteristics of assembly systems." *IIE Transactions* 14(3):204–214.

Williams, J. L. 2005. "Gaussian mixture reduction for tracking multiple maneuvering targets in clutter." Ph.D. thesis, Air Force Institute of Technology, Ohio.

Wu, C. F. J. 1983. "On the convergence properties of the EM algorithm." *Annals of Statistics* 11:55–103.

Yang, J., and Yu, G. 2002. "On the robust single machine scheduling problem." *Journal of Combinatorial Optimization* 6:17–33.

Yao, M., and Chu, W. 2007. "A new approximation algorithm for obtaining the probability distribution function for project completion time." *Computers and Mathematics with Applications* 54(2):282–295.

Index